Probability Theory and Stochastic Modelling

Volume 73

The **Stochastic Modelling and Probability Theory** series is a merger and continuation of Springer's two well established series Stochastic Modelling and Applied Probability and Probability and Its Applications series. It publishes research monographs that make a significant contribution to probability theory or an applications domain in which advanced probability methods are fundamental. Books in this series are expected to follow rigorous mathematical standards, while also displaying the expository quality necessary to make them useful and accessible to advanced students as well as researchers. The series covers all aspects of modern probability theory including

- Gaussian processes
- Markov processes
- Random Fields, point processes and random sets
- Random matrices
- Statistical mechanics and random media
- Stochastic analysis

as well as applications that include (but are not restricted to):

- Branching processes and other models of population growth
- Communications and processing networks
- Computational methods in probability and stochastic processes, including simulation
- Genetics and other stochastic models in biology and the life sciences
- Information theory, signal processing, and image synthesis
- Mathematical economics and finance
- Statistical methods (e.g. empirical processes, MCMC)
- Statistics for stochastic processes
- Stochastic control
- Stochastic models in operations research and stochastic optimization
- Stochastic models in the physical sciences

More information about this series at http://www.springer.com/series/13205

Vidyadhar Mandrekar · Barbara Rüdiger

Stochastic Integration in Banach Spaces

Theory and Applications

Vidyadhar Mandrekar
Department of Statistics and Probability
Michigan State University
East Lansing, MI
USA

Barbara Rüdiger
Department of Mathematics and Informatics
University of Wuppertal
Wuppertal
Germany

ISSN 2199-3130
ISBN 978-3-319-36522-0
DOI 10.1007/978-3-319-12853-5

ISSN 2199-3149 (electronic)
ISBN 978-3-319-12853-5 (eBook)

Mathematics Subject Classification (2010): 60H15, 60H05, 60G57, 60G51, 91G30, 91G80, 60G35, 35B40

Springer Cham Heidelberg New York Dordrecht London

Printed on acid-free paper

Springer is part of Springer Science+Business Media (www.springer.com)

We dedicate this book to the memory of Anatolli Skorokhod and to our teachers in this field G. Kallianpur and Sergio Albeverio

Contents

Chapter 1
Introduction

The study of stochastic differential equations (SDEs) driven by Lévy processes in $\mathbb{R}$ originated in the book by Skorokhod [97]. In view of the Lévy–Itô decomposition, he reduced the problem of studying such SDEs to the analysis of SDEs driven by compensated Poisson random measures (cPrms) and Brownian motion, under a mild restriction [97]. He was aware of the fact that the restriction can be removed. Recently, following initial work of Eberlein and Özkan [27], these SDEs have been found to arise in finance as term structure models for interest rates, volatility in market indices [9] and in the study of flows with applications to pseudo-differential equations [16]. The more general SDEs studied in [36] arise in polymer models [24].

Here, we consider SDEs driven by non-Gaussian Lévy processes, and hence, following Skorokhod, we examine SDEs driven by compensated Poisson random measures. For this, Skorokhod starts by defining an Itô stochastic integral with respect to compensated Poisson random measures with associated variance measure $\lambda \otimes \beta$, where λ denotes the Lebesgue measure and β is a Lévy measure. He defines the stochastic integral of non-anticipating functions with respect to filtering associated with compensated Poisson random measures.

Later, Ikeda and Watanabe in their fundamental book [45] generalized the definition of the Itô integral to compensated Poisson point processes. Here, the associated variance measure is a general measure on $\mathbb{R}$ allowing jumps (as opposed to Lebesgue measure). This means the integral has to be defined with respect to predictable processes [96] which remove Skorokhod's restriction and introduce the interlacing of solutions with respect to compensated Poisson random measures at jump times.

So, in order to generalize Skorokhod's work to infinite-dimensional spaces, one needs to define the Itô stochastic integral of non-anticipating functions taking values in a Banach space with respect to compensated Poisson random measure. In the case where the Banach space is Hilbertian, the definition can be given using the same techniques as in the one-dimensional case and one obtains an Itô isometry for the Itô integral. This was done in the work of Rüdiger [91].

In [53] Kallianpur and Xiong studied SDEs driven by cPrms in multi-Hilbertian spaces with interesting applications to pollution in rivers. Since in multi-Hilbertian

© Springer International Publishing Switzerland 2015
V. Mandrekar and B. Rüdiger, *Stochastic Integration in Banach Spaces*,
Probability Theory and Stochastic Modelling 73, DOI 10.1007/978-3-319-12853-5_1

spaces the boundedness of a set implies its compactness, this theory is, although interesting, rather restrictive.

In view of the fact that a Hilbert space has an inner product and by the validity of the Itô isometry, one can easily extend Skorokhod's technique for SDEs to Hilbert space SDEs. A restricted program in this direction with a predictability assumption and using Ikeda–Watanabe interlacing was carried out in the first book on Hilbert space valued SDEs [5]. Due to its clarity, this book has generated a lot of interest in applying SDEs in specific applied problems.

We have written this monograph with an eye towards applications. Here, we study SDEs in Banach spaces and stochastic partial differential equations (SPDEs), both driven by compensated Poisson random measures. In Chap. 6, we show how SPDEs occur naturally in filtering problems and in applications to finance. If we set the partial differential operator identically equal to zero, then these results reduce to generalized versions of results from [5].

S(P)DEs in Banach spaces arise naturally in different problems as seen, for example, in [12] for applications in finance or, for fluid dynamics, in [29] (see also [14, 30]). In order to describe the theory in this generality we follow Skorokhod's program, first defining stochastic integrals in Banach spaces.

In Chap. 3, we begin by considering the Wiener integral of deterministic functions with values in a separable Banach space. This first appeared in Albeverio and Rüdiger [3], establishing the Lévy–Itô decomposition. The assumption required on the Banach space is that it is of (Rademacher) type 2 (see [42]). This is a condition on the geometry of the Banach space. The Wiener integral of Bochner square-integrable functions can then be defined with respect to any compensated Poisson random measure. However, to study this integral for this class of functions for a particular Poisson random measure with compensator $\lambda \otimes \beta$, we only need inequality (3.1.4), connecting the Wiener integral of a simple function f with respect to the compensated Poisson random measure to the Bochner integral of $\|f\|^2$ with respect to the compensator $\lambda \otimes \beta$. In fact, we prove that the constant K_β in (3.1.4) is independent of β if and only if the space is of type 2 (see [67]).

Motivated by this, we study the Itô integral for non-anticipative processes with respect to compensated Poisson random measures satisfying inequality (3.5.7), which is the analogue of inequality (3.1.4). In fact, using ideas of Rosinski [90], we prove that (3.5.7) is necessary for defining the Itô integral for all processes which are square-integrable with respect to $\lambda \otimes \beta \otimes \mathbb{P}$ (see [67]). Then we show that in M-type 2 spaces, which are defined in [85], our condition (3.5.7) is satisfied. We remark, however, that inequality (3.5.7) is independent of the geometry of the Banach space, i.e. the condition that a Banach space is of M-type 2 is only a sufficient condition for the validity of inequality (3.5.7). For the definition of the Itô integral, we also prove that the space of simple processes is dense in the space of non-anticipative square-integrable functions with respect to $\lambda \otimes \beta \otimes \mathbb{P}$.

In addition, we establish Itô's formula, based on [68], improving the earlier work of [93].

Once these preliminaries are established, we carry out Skorokhod's plan for SDEs in Banach spaces of M-type 2, establishing existence and uniqueness results for SDEs driven by compensated Poisson random measures. The method here goes through for general Banach spaces, provided condition (3.5.7) is satisfied for each Poisson random measure with compensator $\lambda \otimes \beta$. This work first appeared in [67] and was done independently for M-type 2 spaces in [38].

The study of SPDEs driven by Brownian motion was initiated by Pardoux [80] and Krylov and Rozovskii [56], who were motivated by attempts to solve the Zakai equation, which occurs in filtering problems. They studied the strong solutions to the so-called variational problem. The corresponding equation for SPDEs driven by jump Lévy processes are studied in [89]. Their presentation is very clear and is easily accessible to an advanced researcher. The mild solution to SPDEs driven by Brownian motion was originally studied by Da Prato and Zabczyk [18]. For the Brownian motion case, the Pardoux, Krylov–Rozovskii and Da Prato–Zabczyk approach is presented for the interested reader in the book [34]. In [75], the analogue of the Zakai equation with jump processes is studied. However, their approach to the solution is through Malliavin calculus and uses the work of Di Nunno et al. [22]. In Chap. 6, in order to make the presentation self-contained, we show how the Zakai equation with jump processes arises in filtering problems, based on the work in [64]. Even in the Brownian motion case, the approaches to solving filtering problems depend on the structure, as one can see in [101]. Independently of filtering problems, the study of SPDEs driven by cPrms was initiated in [4], in the case where the partial differential operator is the Laplacian, motivated by applications in physics.

The material in Chap. 5 on SPDEs is based on [2], generalizing the work in [4]. We also study the non-Markovian case, due to its applications, see [24]. The solution concept we study is that of a mild solution [83], which is the analogue of that studied in [18, 34, 35] in the case of Brownian motion. In the Markovian case, we study smoothness with respect to initial value in terms of Gateaux differentiability. An improvement of this part was subsequently given in [72] in terms of Fréchét differentiability. We do not include here the fundamental work of [89] on the variational method for Lévy driven SPDEs as including it would mean rewriting their work. We present in Chap. 6 an application of these ideas for the HJM model in finance given in [32]. As one can see, it requires effort to formulate the HJM-equation and to find a proper Hilbert space based on the work in [13, 31]. In order to keep the book self-contained, we present this formulation following the basic work of Filipovic and Tappe [32].

Finally, we study asymptotic properties of the solutions of SPDEs driven by compensated Poisson random measures using the method of Lyapunov functions. This extends the work initiated in [54, 59, 60, 63] for the Brownian motion case. This work is based on Li Wang's thesis [99] and is taken from [69]. Here we use Yosida approximations of mild solutions by strong solutions. One of these was given in [2] and the other is in [65]. It should be noted that the Lyapunov function method studied here is new, even for SDEs driven by compensated Poisson random measures in one dimension. It can be used, for example, to study the exponential stability needed for almost sure stability using the techniques of [70] as in [6] by constructing a Lyapunov function in specific examples. This is also done to study exponential ultimate boundedness for spot rate models driven by Lévy processes [11]. As a consequence, one can prove the positive recurrence property of spot rate models to an interval and use it to make investment decisions. In particular we can easily prove the asymptotic properties of the HJM model given in [71].

In their recent book [84] Peszat and Zabczyk study stochastic partial differential equations driven by a square-integrable martingale taking values in a Hilbert space. Solutions in their case have cádlàg modification as opposed to our case (for cPrm), where the solution is cádlàg. We give general conditions where the semi-group generated by a partial differential operator (PDO) is a pseudo contraction. We also present general theorems on the asymptotic behaviour of solutions using the Lyapunov method. From these results one can obtain more detailed asymptotic behaviour (recurrence) than the existence of invariant measures derived from the behaviour of solutions for deterministic equations. As the latter behaviour is dependent on the semigroups generated by a PDO, we do not have to treat each PDO case separately, which is done in [84]. We also present a Zakai equation to motivate our study.

The organization of this monograph is as follows: In Chap. 3, we study Wiener and Itô integrals with respect to compensated Poisson random measures and Itô's formula after presenting preliminary concepts in Chap. 2. In Chap. 4 we give existence and uniqueness results for SDEs in Banach spaces. SPDEs driven by compensated Poisson random measures are studied in Chap. 5. In Chap. 7, the asymptotic behaviour of the solutions of SPDEs is examined using Lyapunov's method. We show in Chap. 6 how the SPDEs driven by compensated Poisson random measures arise in filtering problems and how our general results can be applied to get the existence of solutions for the HJM model from finance on appropriate function spaces.

We do not give other examples of SPDEs involving partial differential operators such as Navier–Stokes, Reaction–Diffusion etc. because most of these can be worked out as in the book [34] given for Brownian motion as noise. As the results for Lévy driven SPDEs are similar to those given in [34], these examples can be worked out as exercises.

Acknowledgments We would like to thank Professor A.V. Skorokhod who inspired the ideas of this work and Professor S. Albeverio for his influence on applications to SPDEs. We clearly benefitted from discussions with Professor F. Proske, Professor T. Meyer-Brandis and Dr. Li Wang, joint work with whom influenced parts of this book, and Professor Stefan Tappe, who was instrumental in the initial part of the book concerning predictability. In particular, we thank him for carefully reading the initial version and suggesting ideas which led to alternative proofs. It would be amiss not to mention Professor Gawarecki, discussions with whom helped in the organization of the material.

Barbara Rüdiger thanks her daughter Chiara Mastandrea, who is the light of her life.

Chapter 2
Preliminaries

In this chapter, we prepare the required preliminaries.

2.1 The Bochner and the Pettis Integral

Let F be a separable Banach space with dual space F^*. We denote by $\mathcal{B}(F)$ the Borel
σ-algebra of F, which is defined as $\mathcal{B}(F) = \sigma(\mathcal{O})$, where $\mathcal{O}$ denotes the system of
all open sets in F, and $\sigma(\cdot)$ as usual denotes the generated σ-algebra.

For $x \in F$ and $\epsilon > 0$ we denote by $U_\epsilon(x)$ the open ball around x with radius ϵ,
that is

$$U_\epsilon(x) = \{y \in F : \|y - x\| < \epsilon\}.$$

We denote by $\mathcal{U}$ the system of all open balls in F. Furthermore, we denote by $\mathcal{C}$ the
system of all cylinder sets

$$\{x_1^* \in B_1, \ldots, x_n^* \in B_n\}$$

with $n \in \mathbb{N}$, linear functionals $x_1^*, \ldots, x_n^* \in F^*$ and Borel sets $B_1, \ldots, B_n \in \mathcal{B}(\mathbb{R})$.

Proposition 2.1.1 *Suppose the Banach space F is separable. Then we have*

$$\mathcal{B}(F) = \sigma(\mathcal{U}) = \sigma(\mathcal{C}).$$

Proof For any open set $O \in \mathcal{O}$, by Lindelöf's Lemma [1, Lemma 1.1.6] there exist
sequences $(x_n)_{n \in \mathbb{N}} \subset F$ and $(\epsilon_n)_{n \in \mathbb{N}} \subset (0, \infty)$ such that $O = \cup_{n \in \mathbb{N}} U_{\epsilon_n}(x_n) \in \sigma(\mathcal{U})$.
Hence $\mathcal{O} \subset \sigma(\mathcal{U})$, proving $\mathcal{B}(F) \subset \sigma(\mathcal{U})$.

Let $U \in \mathcal{U}$ be an open ball. Then there exist $x \in F$ and $\epsilon > 0$ such that $U = U_\epsilon(x)$.
By [41, Theorem 2.8.5] there exists a sequence $(x_n^*)_{n \in \mathbb{N}} \subset F^*$ such that

$$\|x\| = \sup_{n \in \mathbb{N}} |\langle x_n^*, x \rangle| \quad \text{for all} \quad x \in F.$$

© Springer International Publishing Switzerland 2015
V. Mandrekar and B. Rüdiger, *Stochastic Integration in Banach Spaces*,
Probability Theory and Stochastic Modelling 73, DOI 10.1007/978-3-319-12853-5_2

Therefore, we obtain

$$U = \{y \in F : \|y - x\| < \epsilon\} = \{y \in F : \sup_{n \in \mathbb{N}} |\langle x_n^\star, y - x \rangle| < \epsilon\}$$

$$= \cap_{n \in \mathbb{N}} \{y \in F : |\langle x_n^\star, y \rangle - \langle x_n^\star, x \rangle| < \epsilon\} = \cap_{n \in \mathbb{N}} \{x_n^\star \in U_\epsilon(\langle x_n^\star, x \rangle)\} \in \sigma(C).$$

This shows $\mathcal{U} \subset \sigma(C)$, and hence $\sigma(\mathcal{U}) \subset \sigma(C)$.

For any $n \in \mathbb{N}$ and linear functionals $x_1^\star, \ldots, x_n^\star \in F^\star$ the mapping

$$(x_1^\star, \ldots, x_n^\star) : E \to \mathbb{R}^n$$

is continuous. Therefore, we have $C \subset \mathcal{B}(F)$, and hence $\sigma(C) \subset \mathcal{B}(F)$.   □

Let $(\Omega, \mathcal{F}, \mu)$ be a σ-finite measure space. Then we call a function $f : \Omega \to F$ *measurable* if it is $\mathcal{F}/\mathcal{B}(F)$-measurable, and we call f *weakly measurable* if for each $x^* \in F^*$ the scalar valued function $\langle x^*, f \rangle$ is measurable. If $(\Omega, \mathcal{F}, \mu)$ is a probability space, that is $\mu(\Omega) = 1$, we will also call a measurable function $f : \Omega \to F$ a *random variable*.

A measurable function $f : \Omega \to F$ is called *simple* (or *elementary*) if there exist a positive integer $n \in \mathbb{N}$, elements $x_1, \ldots, x_n \in F$ and sets $A_1, \ldots, A_n \in \mathcal{F}$ with $\mu(A_i) < \infty$ for $i = 1, \ldots, n$ such that

$$f = \sum_{i=1}^n x_i \mathbb{1}_{A_i}. \tag{2.1.1}$$

We denote by $\mathcal{E}(F) = \mathcal{E}(\Omega, \mathcal{F}, \mu; F)$ the linear space of all elementary functions.

A function $f : \Omega \to F$ is called *strongly measurable* if there exists a sequence $(f_n)_{n \in \mathbb{N}} \subset \mathcal{E}(F)$ of simple functions such that $f_n(\omega) \to f(\omega)$ for all $\omega \in \Omega$.

As the Banach space F is separable, a function $f : \Omega \to E$ is weakly measurable if and only if it is measurable if and only if it is strongly measurable.

For $p \geq 1$ we denote by $L^p(\Omega, \mathcal{F}, \mu; F)$ the linear space consisting of all measurable functions $f : \Omega \to F$ such that

$$\left(\int_\Omega \|f\|^p d\mu \right)^{1/p} < \infty.$$

Identifying all measurable functions which coincide μ-almost everywhere, the linear space $L^p(\Omega, \mathcal{F}, \mu; F)$ is a Banach space. We shall also use the abbreviation $L^p(F)$, or $L^p(\mathcal{F}; F)$, when there is no chance of ambiguity.

Let us now provide the definition of the Bochner integral. For a simple function $f \in \mathcal{E}(F)$ of the form (2.1.1) we define the Bochner integral as

$$\int_\Omega f d\mu := \sum_{i=1}^n x_i \mu(A_i).$$

Then we have

$$\left\| \int_\Omega f d\mu \right\| = \left\| \sum_{i=1}^n x_i \mu(A_i) \right\| \le \sum_{i=1}^n \|x_i\| \mu(A_i) = \int_\Omega \|f\| d\mu.$$

Therefore, the Bochner integral defines a continuous linear operator

$$\mathcal{E}(F) \to F, \quad f \mapsto \int_\Omega f d\mu. \tag{2.1.2}$$

Lemma 2.1.2 *For each $p \ge 1$ the linear space $\mathcal{E}(F)$ is dense in $L^p(F)$.*

Consequently, the integral operator (2.1.2) has a unique extension

$$L^1(F) \to F, \quad f \mapsto \int_\Omega f d\mu, \tag{2.1.3}$$

which we also call the *Bochner integral*, and we have the estimate

$$\left\| \int_\Omega f d\mu \right\| \le \int_\Omega \|f\| d\mu \quad \text{for all } f \in L^1(F). \tag{2.1.4}$$

If the measure space $(\Omega, \mathcal{F}, \mu)$ is finite, that is, $\mu(\Omega) < \infty$, then, by the Cauchy–Schwarz inequality, for each $p \ge 1$ the integral operator (2.1.2) has a unique continuous extension

$$L^p(F) \to F, \quad f \mapsto \int_\Omega f d\mu.$$

For a function $f \in L^1(F)$ and a subset $A \in \mathcal{F}$ we set

$$\int_A f d\mu := \int_\Omega f \mathbb{1}_A d\mu.$$

If $(\Omega, \mathcal{F}, \mu)$ is a probability space, that is, $\mu(\Omega) = 1$, for $f \in L^1(F)$ we set

$$\mathbb{E}[f] := \int_\Omega f d\mu.$$

If the measure space is given by $(\Omega, \mathcal{F}, \mu) = (\mathbb{R}_+, \mathcal{B}(\mathbb{R}_+), \lambda)$ where λ denotes the Lebesgue measure, then for each measurable function $f : \mathbb{R}_+ \to F$ with

$$\int_0^t \|f(s)\| ds < \infty \quad \text{for all } t \ge 0$$

we define the function

$$\int_0^t f(s) ds := \int_{(0,t]} f d\lambda, \quad t \ge 0.$$

Proposition 2.1.3 *The function* $\mathbb{R}_+ \to F$, $t \mapsto \int_0^t f(s)ds$, *is continuous.*

Now let $(\Omega_i, \mathcal{F}_i)$, $i = 1, 2$, be two measure spaces and let μ_2 be a σ-finite measure on Ω_2. We set

$$(\Omega, \mathcal{F}) = (\Omega_1 \times \Omega_2, \mathcal{F}_1 \otimes \mathcal{F}_2).$$

Proposition 2.1.4 *Let* $f : \Omega \to F$ *be measurable such that* $\omega_2 \mapsto f(\omega_1, \omega_2) \in L^1(\Omega_2; F)$ *for all* $\omega_1 \in \Omega_1$. *Then the mapping*

$$\Omega_1 \to F, \quad \omega_1 \mapsto \int_{\Omega_2} f(\omega_1, \omega_2)\mu_2(d\omega_2) \tag{2.1.5}$$

is measurable.

Now let G be another separable Banach space. Recall that for a closed linear operator $A : \mathcal{D}(A) \subset F \to G$ the domain $\mathcal{D}(A)$ equipped with the graph norm $\|\|x\|\|_{\mathcal{D}(A)} = \|x\| + \|Ax\|$ is also a Banach space. Using the closed graph theorem, we can prove the following result:

Proposition 2.1.5 *Let* $A : \mathcal{D}(A) \subset F \to G$ *be a closed operator and let* $f \in L^1(\mathcal{D}(A))$ *be a function. Then we have* $f \in L^1(F)$, $Af \in L^1(G)$ *and*

$$A \int_\Omega f d\mu = \int_\Omega A f d\mu. \tag{2.1.6}$$

Recall that a system $\mathcal{S} \subset \mathcal{F}$ is called a *semiring* if:

1. $\emptyset \in \mathcal{S}$;
2. For all $A, B \in \mathcal{S}$ we have $A \cap B \in \mathcal{S}$;
3. For all $A, B \in \mathcal{S}$ there exist $n \in \mathbb{N}$ and disjoint sets $C_1, \ldots, C_n \in \mathcal{S}$ such that $A \setminus B = \bigcup_{i=1}^n C_i$.

We denote by $\Sigma(F) = \Sigma(\mathcal{S}; F)$ the linear space of all simple functions $f : \Omega \to F$ of the form (2.1.1) with disjoint sets $A_1, \ldots, A_n \in \mathcal{S}$.

Proposition 2.1.6 *Let S be a semiring such that* $\mathcal{F} = \sigma(\mathcal{S})$. *Then, for all $p \geq 1$ the linear space $\Sigma(\mathcal{S}; F)$ is dense in $L^p(F)$.*

Proof Let $p \geq 1$ and $f \in L^p(F)$ be arbitrary. By Lemma 2.1.2 we may assume that f is of the form (2.1.1) with $n \in \mathbb{N}$, elements $x_1, \ldots, x_n \in F$ and sets $A_1, \ldots, A_n \in \mathcal{F}$ such that $\mu(A_i) < \infty$ for $i = 1, \ldots, n$. Let us first assume that the measure μ is finite. Note that the σ-algebra $\mathcal{F}$ is generated by the algebra

$$\mathcal{A} = \{B_1 \cup \ldots \cup B_p : p \in \mathbb{N} \text{ and } B_1, \ldots, B_p \in \mathcal{S} \text{ are disjoint}\}.$$

Since the measure μ is finite [10, Theorem 5.7] applies and yields for each $m \in \mathbb{N}$ and each $i = 1, \ldots, n$ the existence of a set $B_i^m \in \mathcal{A}$ such that

$$\mu(A_i \Delta B_i^m) < \frac{1}{n^p m \|x_i\|^p}. \tag{2.1.7}$$

We define the sequence $(f_m)_{m \in \mathbb{N}} \subset \mathcal{E}(F)$ of simple functions as

$$f_m := \sum_{i=1}^{n} x_i \mathbb{1}_{B_i^m}, \quad m \in \mathbb{N}.$$

Note that, due to the third property of the semiring $\mathcal{S}$, we have $(f_m)_{m \in \mathbb{N}} \subset \Sigma(\mathcal{S}; F)$. It remains to prove that $f_m \to f$ in $L^p(F)$. By (2.1.7), for each $m \in \mathbb{N}$ we obtain

$$\int_\Omega \|f - f_m\|^p d\mu = \int_\Omega \left\| \sum_{i=1}^{n} x_i (\mathbb{1}_{A_i} - \mathbb{1}_{B_i^m}) \right\|^p d\mu \le \int_\Omega \left(\sum_{i=1}^{n} \|x_i\| \mathbb{1}_{A_i \Delta B_i^m} \right)^p d\mu$$

$$\le n^{p-1} \int_\Omega \sum_{i=1}^{n} \|x_i\|^p \mathbb{1}_{A_i \Delta B_i^m} d\mu = n^{p-1} \sum_{i=1}^{n} \|x_i\|^p \mu(A_i \Delta B_i^m)$$

$$< n^{p-1} \sum_{i=1}^{n} \|x_i\|^p \frac{1}{n^p m \|x_i\|^p} = n^{p-1} \frac{n}{n^p m} = \frac{1}{m},$$

which completes the proof for the case when the measure μ is finite or when f has finite support.

Suppose f does not have finite support. Then for all $\epsilon > 0$ there is an f_ϵ with finite support such that $\int_\Omega \|f - f_\epsilon\|^p d\mu < \infty$. For all $\epsilon > 0$ and all $N \in \mathbb{N}$ there is an $m_\epsilon > N$ such that $f_{m_\epsilon} \in \Sigma(F)$ and

$$\int_\Omega \|f - f_{m_\epsilon}\|^p d\mu \le 2^p \int_\Omega \|f - f_\epsilon\|^p d\mu + 2^p \int_\Omega \|f_\epsilon - f_{m_\epsilon}\|^p d\mu \le 2^p \epsilon + \frac{2^p}{m_\epsilon},$$

which completes the proof. $\square$

Let us now recall the definition of the Pettis integral. The following result is known as Dunford's lemma.

Lemma 2.1.7 *Let* $f : \Omega \to F$ *be a weakly measurable function such that* $\langle x^*, f \rangle \in L^1(\mathbb{R})$ *for all* $x^* \in F^*$. *Then the linear operator*

$$x_f^{**} : F^* \to \mathbb{R}, \quad \langle x_f^{**}, x^* \rangle := \int_\Omega \langle x^*, f \rangle d\mu$$

is continuous.

Let $f : \Omega \to F$ be a weakly measurable function such that $\langle x^*, f \rangle \in L^1(\mathbb{R})$ for all $x^* \in F^*$. By Dunford's lemma we have $x_f^{**} \in F^{**}$ and

$$\text{(D-)} \int_\Omega f d\mu := x_f^{**}$$

is the so-called *Dunford integral* of f. If we have $x_f^{**} \in F$ (which is in particular true if the Banach space F is reflexive), then we say the function f is *Pettis integrable*, and the *Pettis integral* of f is given by

$$\text{(P-)} \int_\Omega f d\mu := x_f^{**}. \tag{2.1.8}$$

Proposition 2.1.8 *Suppose the separable Banach space F is reflexive. Then each function $f \in L^1(F)$ is Pettis integrable and we have*

$$\int_\Omega f d\mu = \text{(P-)} \int_\Omega f d\mu \quad \text{for all } f \in L^1(F). \tag{2.1.9}$$

Moreover, for each $x^* \in F^*$ we have

$$\left\langle x^*, \int_\Omega f d\mu \right\rangle = \int_\Omega \langle x^*, f \rangle d\mu \quad \text{for all } f \in L^1(F). \tag{2.1.10}$$

Proof The identity (2.1.10) follows from Proposition 2.1.5 and (2.1.9) is a consequence of (2.1.10). □

2.2 Stochastic Processes in Banach Spaces

Let $(\Omega, \mathcal{F}, (\mathcal{F}_t)_{t \geq 0}, \mathbb{P})$ be a filtered probability space. The filtration $(\mathcal{F}_t)_{t \geq 0}$ satisfies the *usual conditions* if:

1. It is right-continuous, i.e., we have $\mathcal{F}_t = \bigcap_{s > t} \mathcal{F}_s$ for all $t \geq 0$;
2. It is complete, i.e., we have $\mathcal{N} \subset \mathcal{F}_t, \forall t \in \mathbb{R}_+$, where $\mathcal{N}$ denotes the collection of $\mathbb{P}$-null sets of $\mathcal{F}$.

In the sequel, we shall always assume that the usual conditions are satisfied.

Let F be a Banach space and let $X = (X_t)_{t \in I}$ be an F-valued process with index set $I \subset \mathbb{R}_+$. In the sequel, we will always have $I = \mathbb{R}_+$ or $I = [0, T]$ for some $T > 0$.

The process X is called $(\mathcal{F}_t)$-*adapted* (in short, *adapted*) if for each $t \in I$ the random variable X_t is $\mathcal{F}_t$-measurable.

The process X is *continuous* if all its paths $t \leadsto X_t(\omega)$ are continuous. The process X is *càdlàg* if all its paths are right-continuous with left-hand limits. If X is càdlàg, we define the left-continuous process X_{t-} as

$$X_{t-} := \begin{cases} X_0, & t = 0 \\ \lim_{s \uparrow\uparrow t} X_s, & t > 0 \end{cases}$$

and the jumps ΔX as

$$\Delta X_t := X_t - X_{t-}, \quad t \geq 0.$$

For $T > 0$ we define the system of sets

$$\mathcal{G}_T = \{A \times \{0\} : A \in \mathcal{F}_0\} \cup \{A \times (s, t] : 0 \leq s < t \leq T : A \in \mathcal{F}_s\}$$

and the predictable σ-algebra $\mathcal{P}_T = \sigma(\mathcal{G}_T)$. An F-valued process $X = (X_t)_{t \in [0,T]}$ is called *predictable* if it is $\mathcal{P}_T$-measurable.

We call the sigma algebra generated by all càdlàg processes the "optional sigma algebra" and denote it by $\tilde{\mathcal{P}}_T$.

Lemma 2.2.1 *Every left-continuous, adapted process $X = (X_t)_{t \in [0,T]}$ is predictable.*

Proof For each $n \in \mathbb{N}$ we define the process $X^n = (X_t^n)_{t \in [0,T]}$ as

$$X^n := X_0 \mathbb{1}_{\{0\}} + \sum_{k=1}^{2^n} X_{\frac{T(k-1)}{2^n}} \mathbb{1}_{(\frac{T(k-1)}{2^n}, \frac{Tk}{2^n}]}.$$

Then each X^n is $\mathcal{G}_T$-measurable and, by the left-continuity of X, we have $X^n \to X$ everywhere. Therefore, the process X is predictable. $\quad\square$

Let $\mathcal{H}_T$ be the system of sets

$$\mathcal{H}_T = \{X^{-1}(B) : X = (X_t)_{t \in [0,T]} \text{ is left-continuous, adapted and } B \in \mathcal{B}(F)\}.$$

Proposition 2.2.2 *We have $\mathcal{P}_T = \sigma(\mathcal{H}_T)$, that is, the predictable σ-algebra $\mathcal{P}_T$ is the smallest σ-algebra generated by all left-continuous adapted processes.*

Proof By Lemma 2.2.1 every left-continuous adapted process is predictable, that is, we have $\mathcal{H}_T \subset \mathcal{P}_T$, and hence $\sigma(\mathcal{H}_T) \subset \mathcal{P}_T$. Conversely, every process $X = (X_t)_{t \in [0,T]}$ of the form $X = x\mathbb{1}_A$ with $x \in F$ and $A \in \mathcal{G}_T$ is left-continuous and adapted. Therefore, we have $\mathcal{G}_T \subset \mathcal{H}_T$, and hence $\mathcal{P}_T \subset \sigma(\mathcal{H}_T)$. $\quad\square$

A process $X = (X_t)_{t \geq 0}$ is called *predictable* if for each $T \geq 0$ the restriction $X|_{\Omega \times [0,T]}$ is predictable.

A process $X = (X_t)_{t \in I}$ is called *progressively measurable* if for each $t \in I$ the restriction $X|_{\Omega \times [0,t]}$ is $\mathcal{F}_t \otimes \mathcal{B}([0, t])$-measurable. Note that a progressively measurable process X is also adapted.

Proposition 2.2.3 *Let $T > 0$ be arbitrary. The following statements are valid:*

1. *Every predictable process $X = (X_t)_{t\in I}$ is progressively measurable.*
2. *Every right-continuous adapted process $X = (X_t)_{t\in I}$ is progressively measurable.*

Proof Let $T \in I$ be arbitrary. We have $\mathcal{G}_T \subset \mathcal{F}_T \otimes \mathcal{B}([0, T])$, and hence $\mathcal{P}_T \subset \mathcal{F}_T \otimes \mathcal{B}([0, T])$. Therefore, every predictable process is progressively measurable.

Suppose $X = (X_t)_{t\in[0,T]}$ is right-continuous and adapted. For each $n \in \mathbb{N}$ we define the process $X^n = (X_t^n)_{t\in[0,T]}$ as

$$X^n := X_0 \mathbb{1}_{\{0\}} + \sum_{k=1}^{2^n} X_{\frac{Tk}{2^n}} \mathbb{1}_{(\frac{T(k-1)}{2^n}, \frac{Tk}{2^n}]}.$$

Then each X^n is $\mathcal{F}_T \otimes \mathcal{B}([0, T])$-measurable and, by the right-continuity of X, we have $X^n \to X$ everywhere. Therefore, the process X is $\mathcal{F}_T \otimes \mathcal{B}([0, T])$-measurable, and consequently, every right-continuous adapted process is progressively measurable. □

For $T > 0$ we denote by $\mathcal{K}_T^1(F)$ the linear space of all progressively measurable processes $X = (X_t)_{t\in[0,T]}$ such that

$$\mathbb{P}\left(\int_0^T \|X_s\| ds < \infty \right) = 1.$$

For each $X \in \mathcal{K}_T^1(F)$ we define the pathwise Bochner integral

$$\int_0^t X_s ds, \quad t \in [0, T]. \tag{2.2.1}$$

Lemma 2.2.4 *For each $X \in \mathcal{K}_T^1(F)$ the integral process* (2.2.1) *is continuous and adapted.*

Proof This is a consequence of Propositions 2.1.3 and 2.1.4. □

We denote by $\mathcal{K}_\infty^1(F)$ the linear space of all progressively measurable processes $X = (X_t)_{t\geq 0}$ such that for all $T > 0$ the restriction $X|_{\Omega \times [0,T]}$ belongs to $\mathcal{K}_T^1(F)$. For each $X \in \mathcal{K}_T^1(F)$ we define the pathwise Bochner integral

$$\int_0^t X_s ds, \quad t \geq 0$$

which, according to Lemma 2.2.4, is a continuous adapted process.

Let $X = (X_t)_{t\in I}$ and $Y = (Y_t)_{t\in I}$ be two processes. Then Y is called a *version* (or a *modification*) of X if

$$\mathbb{P}(X_t = Y_t) = 1 \quad \text{for all } t \in I.$$

The processes X and Y are called *indistinguishable* if

$$\mathbb{P}\left(\bigcap_{t \in I}\{X_t = Y_t\}\right) = 1.$$

The processes X and Y are called *independent* if for all $n, m \in \mathbb{N}$ and all $t_1 \leq \ldots \leq t_n$, $s_1, \ldots, s_m$ with $t_1, \ldots, t_n, s_1, \ldots, s_m \in I$ the random vectors $(X_{t_1}, \ldots, X_{t_n})$ and $(Y_{s_1}, \ldots, Y_{s_m})$ are independent.

2.3 Martingales in Banach Spaces

Let $(\Omega, \mathcal{F}, \mathbb{P})$ be a complete probability space and let F be a separable Banach space.

Proposition 2.3.1 *Let $X \in L^1(\mathcal{F}; F)$ be a random variable and let $\mathcal{C} \subset \mathcal{F}$ be a sub σ-algebra. Then, there exists a unique random variable $Z \in L^1(\mathcal{C}; F)$ such that*

$$\mathbb{E}[X\mathbb{1}_C] = \mathbb{E}[Z\mathbb{1}_C] \quad \text{for all } C \in \mathcal{C}.$$

Proof See [18, Proposition 1.10]. $\square$

The random variable Z is denoted by $\mathbb{E}[X \mid \mathcal{C}]$ and is called the *conditional expectation of X given $\mathcal{C}$*. Furthermore, [18, Proposition 1.10] yields that

$$\|\mathbb{E}[X \mid \mathcal{C}]\| \leq \mathbb{E}[\|X\| \mid \mathcal{C}] \quad \text{for all } X \in L^1(\mathcal{F}; F). \tag{2.3.1}$$

Taking into account the Cauchy–Schwarz inequality, for every $p \geq 1$ we may consider the conditional expectation as a continuous linear operator

$$L^p(\mathcal{F}; F) \to L^p(\mathcal{C}; F), \quad X \mapsto \mathbb{E}[X \mid \mathcal{C}].$$

Now let the probability space $(\Omega, \mathcal{F}, \mathbb{P})$ be equipped with a filtration $(\mathcal{F}_t)_{t \geq 0}$ satisfying the usual conditions. Let $I \subset \mathbb{R}_+$ be an index set such that $I = \mathbb{R}_+$ or $I = [0, T]$ for some $T > 0$.

Definition 2.3.2 An F-valued adapted process $(M_t)_{t \in I}$ is called a *martingale* if:

1. We have $\mathbb{E}[\|M_t\|] < \infty$ for all $t \in I$;
2. For all $s, t \in I$ with $s \leq t$ we have $\mathbb{E}[M_t \mid \mathcal{F}_s] = M_s$ almost surely.

Recall that a mapping $\tau : \Omega \to \overline{\mathbb{R}}_+ = [0, \infty]$ is called a *stopping time* if we have $\{\tau \leq t\} \in \mathcal{F}_t$ for all $t \geq 0$. For a stopping time τ and an F-valued process $X = (X_t)_{t \in I}$ the stopped process X^τ is defined as $X_t^\tau := X_{t \wedge \tau}$ for $t \in I$. Note that for a progressively measurable process X and a stopping time τ the stopped process X^τ is progressively measurable, too.

An F-valued process $M = (M_t)_{t \in I}$ is called a *local martingale* if there exists a sequence $(\tau_n)_{n \in \mathbb{N}}$ of stopping times with $\tau_n \uparrow \infty$ almost surely such that for each $n \in \mathbb{N}$ the stopped process M^{τ_n} is a martingale.

Lemma 2.3.3 *An F-valued, adapted process* $M = (M_t)_{t \in I}$ *with* $\mathbb{E}[\|M_t\|] < \infty$ *for all* $t \in I$ *is a martingale if and only if for each* $x^* \in F^*$ *the real-valued process* $\langle x^*, M \rangle$ *is a martingale.*

Proof If M is a martingale, then clearly for each $x^* \in F^*$ the process $\langle x^*, M \rangle$ is a martingale, too. Now suppose that for each $x^* \in F^*$ the process $\langle x^*, M \rangle$ is a martingale. By [41, Theorem 2.8.5] there exists a sequence $(x_n^*)_{n \in \mathbb{N}} \subset F^*$ such that

$$\|x\| = \sup_{n \in \mathbb{N}} |\langle x_n^*, x \rangle| \quad \text{for all } x \in F.$$

Let $s, t \in I$ with $s \leq t$ be arbitrary. There exists a set $\Omega_0 \in \mathcal{F}$ with $\mathbb{P}(\Omega_0) = 1$ such that

$$\mathbb{E}[\langle x_n^*, X_t \rangle \mid \mathcal{F}_s](\omega) = \langle x_n^*, X_s(\omega) \rangle \quad \text{for all } \omega \in \Omega_0 \text{ and } n \in \mathbb{N}.$$

Therefore, for all $\omega \in \Omega_0$ we obtain

$$\|\mathbb{E}[X_t \mid \mathcal{F}_s](\omega) - X_s(\omega)\| = \sup_{n \in \mathbb{N}} |\langle x_n^*, \mathbb{E}[X_t \mid \mathcal{F}_s](\omega) - X_s(\omega) \rangle|$$
$$= \sup_{n \in \mathbb{N}} |\mathbb{E}[\langle x_n^*, X_t \rangle \mid \mathcal{F}_s](\omega) - \langle x_n^*, X_s(\omega) \rangle| = 0,$$

finishing the proof.   □

Theorem 2.3.4 *Every F-valued martingale* $M = (M_t)_{t \in I}$ *has a càdlàg version.*

Proof Since F is separable, we look at $\langle x^*, M_t \rangle$ for x^* in a countable determining set. Then by the one-dimensional result the theorem follows.   □

The following two estimates are known as Doob's L^p-inequalities.

Theorem 2.3.5 *Let* $M = (M_t)_{t \in I}$ *be an F-valued martingale. Then, the following statements are valid:*

1. *For all* $p \geq 1$ *and* $\lambda > 0$ *we have*

$$\mathbb{P}\left(\sup_{t \in [0,T]} \|M_t\| \geq \lambda \right) \leq \frac{1}{\lambda^p} \mathbb{E}[\|M_T\|^p]. \tag{2.3.2}$$

For every $p > 1$ *we have*

$$\mathbb{E}\left[\sup_{t \in [0,T]} \|M_t\|^p \right] \leq \left(\frac{p}{p-1} \right)^p \mathbb{E}[\|M_T\|^p]. \tag{2.3.3}$$

Proof The real-valued process $\|M\|$ is a non-negative submartingale. Indeed, for all $s, t \in I$ with $s \leq t$ we have, by using (2.3.1),

$$\|M_s\| = \|\mathbb{E}[M_t \mid \mathcal{F}_s]\| \leq \mathbb{E}[\|M_t\| \mid \mathcal{F}_s].$$

Thus, the inequalities (2.3.2) and (2.3.3) follow from [88, Theorem II.1.7]. $\qquad\square$

Fix a finite time horizon $T > 0$. For $p \geq 1$ we denote by $\mathcal{M}_T^p(E)$ the linear space of all E-valued martingales $M = (M_t)_{t \in [0,T]}$ with $\mathbb{E}[\|M_T\|^p] < \infty$. Identifying indistinguishable processes, a norm on $\mathcal{M}_T^p(E)$ is given by

$$\|M\|_{\mathcal{M}_T^p} = \mathbb{E}[\|M_T\|^p]^{1/p}.$$

Lemma 2.3.6 *For each $p \geq 1$ the normed space $\mathcal{M}_T^p(E)$ is a Banach space.*

Proof Let $p \geq 1$ be arbitrary. The linear mapping

$$\phi : (\mathcal{M}_T^p(E), \|\cdot\|) \to L^p(\mathcal{F}_T; E), \quad \phi(M) = M_T$$

is in particular an isometry. It is also surjective, because for any $X \in L^p(\mathcal{F}_T; E)$ we have $\phi^{-1}(X) = M$ with $M_t = \mathbb{E}[X_T \mid \mathcal{F}_t]$ for $t \in [0, T]$. Therefore, ϕ is an isometric isomorphism, proving the completeness of $\mathcal{M}_T^p(E)$. $\qquad\square$

2.4 Poisson Random Measures

Let $(\Omega, \mathcal{F}, (\mathcal{F}_t)_{t \geq 0}, \mathbb{P})$ be a filtered probability space and let $(E, \mathcal{E})$ be a Blackwell space. We set $\mathcal{X} = \mathbb{R}_+ \times E$.

We recall here the definition of a Blackwell space (Definition 24, Chap. 3) [20].

Definition 2.4.1 A measurable space $(E, \mathcal{E})$ is a Blackwell space if the associated Hausdorff space $(E, \mathcal{E})$ is Souslin.

Remark 2.4.2 For simplicity, as a particular case, we might take $(E, \mathcal{E})$ to be a complete separable metric space, since in all applications considered in this monograph $(E, \mathcal{E})$ is the space which marks the jumps of a Lévy process. However, we shall state the results of this section in full generality. We remark that a Blackwell space is in particular countably generated and that the σ-algebra $\mathcal{B}(\mathcal{X})$ is generated by the semiring

$$\mathcal{S} = \{\{0\} \times B : B \in \mathcal{E}\} \cup \{(s, t] \times B : 0 \leq s < t \text{ and } B \in \mathcal{E}\}.$$

Definition 2.4.3 A *random measure* on $\mathcal{X}$ is a family $N = \{N(\omega; dt, dx) : \omega \in \Omega\}$ of measures on $(\mathcal{X}, \mathcal{B}(\mathcal{X}))$ satisfying $N(\omega; \{0\} \times E) = 0$ for all $\omega \in \Omega$.

Definition 2.4.4 An *integer-valued random measure* is a random measure N that satisfies:

- $N(\omega; \{t\} \times E) \leq 1$ for all $\omega \in \Omega$;
- $N(\omega; A) \in \overline{\mathbb{N}}_0$ for all $\omega \in \Omega$ and all $A \in \mathcal{B}(\mathcal{X})$;
- N is optional and $\tilde{\mathcal{P}}$-σ-finite, where $\tilde{\mathcal{P}}$ is the σ-field of optional sets.

The third condition is purely technical and ensures that for each progressively measurable process $f : \Omega \times E \times \mathbb{R}_+ \to F$ with

$$\mathbb{P}\left(\int_0^t \int_E \|f(s,x)\| N(ds, dx) < \infty \right) = 1 \quad \text{for all } t > 0$$

the integral process is adapted.

Definition 2.4.5 A *Poisson random measure* on $\mathcal{X}$, relative to $(\mathcal{F}_t)_{t \geq 0}$, is an integer-valued random measure N such that:

1. There exists a σ-finite measure β on E such that

$$\mathbb{E}[N(A)] = (\lambda \otimes \beta)(A), \quad A \in \mathcal{B}(\mathcal{X}).$$

2. For every $s \in \mathbb{R}_+$ and every $A \in \mathcal{B}(\mathcal{X})$ such that $A \subset (s, \infty) \times E$ and $\mathbb{E}[N(A)] < \infty$, the random variable $N(A)$ is independent of $\mathcal{F}_s$.

For a Poisson random measure N the measure $\nu(dt, dx) := dt \otimes \beta(dx)$ is called the *compensator* (or the *intensity measure*) of N. The next result explains the terminology "Poisson random measure".

Theorem 2.4.6 *Let N be a Poisson random measure with compensator ν, and let $A_1, \ldots, A_n \in \mathcal{B}(\mathcal{X})$ be disjoint subsets for some $n \in \mathbb{N}$ with $\nu(A_i) < \infty$ for all $i = 1, \ldots, n$. Then the random variables $N(A_i)$, $i = 1, \ldots, n$ are independent and have a Poisson distribution with mean $\nu(A_i)$.*

Proof The statement follows from [48, Theorem II.4.8]. □

Let N be a Poisson random measure. We call $q(dt, dx) = N(dt, dx) - \nu(dt, dx)$ the *compensated Poisson random measure* associated to N, and ν is also called the *compensator* (or the *intensity measure*) of q. Note that for each set $A \in \mathcal{B}(\mathcal{X})$ with $\nu(A) < \infty$ we have, since $N(A) \sim \text{Pois}(\nu(A))$ according to Theorem 2.4.6,

$$\mathbb{E}[q(A)] = 0 \quad \text{and} \quad \mathbb{E}[q(A)^2] = \nu(A).$$

Lemma 2.4.7 *For each $A \in \mathcal{B}(\mathbb{R}_+)$ and each $B \in \mathcal{E}$ with $\beta(B) < \infty$, the process $M = (M_t)_{t \geq 0}$ given by*

$$M_t = q((0, t] \cap A \times B), \quad t \geq 0$$

is an $(\mathcal{F}_t)$-martingale.

Proof Let $0 \le s \le t < \infty$ be arbitrary. Since $N((s, t] \cap A \times B)$ is independent of $\mathcal{F}_s$ (see Definition 2.4.5) and has distribution $\text{Pois}(\lambda((s, t] \cap A)\beta(B))$ by Theorem 2.4.6, we obtain

$$
\begin{aligned}
\mathbb{E}[M_t - M_s \mid \mathcal{F}_s] &= \mathbb{E}[q((s, t] \cap A \times B) \mid \mathcal{F}_s] \\
&= \mathbb{E}[N((s, t] \cap A \times B) \mid \mathcal{F}_s] - \lambda((s, t] \cap A)\beta(B) \\
&= \mathbb{E}[N((s, t] \cap A \times B)] - \lambda((s, t] \cap A)\beta(B) = 0,
\end{aligned}
$$

establishing the proof.   □

Lemma 2.4.8 *For each $s \ge 0$, each $A \in \mathcal{B}(\mathbb{R}_+)$ with $A \subset (s, \infty)$, each $F \in \mathcal{F}_s$ and each $B \in \mathcal{E}$ with $\beta(B) < \infty$ the process $M = (M_t)_{t \ge 0}$ given by*

$$
M_t = \mathbb{1}_F q((0, t] \cap A \times B), \quad t \ge 0
$$

is an $(\mathcal{F}_t)$-martingale.

Proof By Lemma 2.4.7, the process

$$
N_t = q((0, t] \cap A \times B), \quad t \in [0, T]
$$

is a martingale. We shall now prove that $\mathbb{1}_F N$ is also a martingale, which will finish the proof. Let $0 \le u \le t$ be arbitrary. If $u \le s$, then we have

$$
\mathbb{E}[\mathbb{1}_F N_t \mid \mathcal{F}_u] = \mathbb{E}[\mathbb{1}_F \mathbb{E}[N_t \mid \mathcal{F}_s]] \mid \mathcal{F}_u] = \mathbb{E}[\mathbb{1}_F N_s \mid \mathcal{F}_u] = 0 = \mathbb{1}_F N_u,
$$

and for $u > s$ we obtain

$$
\mathbb{E}[\mathbb{1}_F N_t \mid \mathcal{F}_u] = \mathbb{1}_F \mathbb{E}[N_t \mid \mathcal{F}_u] = \mathbb{1}_F N_u,
$$

showing that $\mathbb{1}_F N$ is a martingale.   □

We set $\mathcal{L}^1_\beta(F) := L^1(\mathcal{X}, \mathcal{B}(\mathcal{X}), \lambda \otimes \beta; F)$.

Theorem 2.4.9 *The linear space $\Sigma(F) = \Sigma(\mathcal{S}; F)$ is dense in $\mathcal{L}^1_\beta(F)$.*

Proof This follows from Remark 2.4.2 and by Proposition 2.1.6.   □

Remark 2.4.10 Any function $f \in \Sigma(F)$ is of the form

$$
f(t, x) = \sum_{k=1}^{n} \sum_{l=1}^{m} a_{k,l} \mathbb{1}_{A_{k,l}}(x) \mathbb{1}_{(t_{k-1}, t_k]}(t) \tag{2.4.1}
$$

for $n, m \in \mathbb{N}$ with:

- elements $a_{k,l} \in F$ for $k = 1, \dots, n$ and $l = 1, \dots, m$;
- time points $0 \le t_0 \le \dots t_n < \infty$;

- sets $A_{k,l} \in \mathcal{E}$ with $\beta(A_{k,l}) < \infty$ for $k = 1, \ldots, n$ and $l = 1, \ldots, m$ such that the product sets $A_{k,l} \times (t_{k-1}, t_k]$ are mutually disjoint.

Definition 2.4.11 For every $f \in \Sigma(F)$ of the form (2.4.1) we define the *Wiener integral* of f with respect to q as

$$\iint_{\mathcal{X}} f(t,x)q(dt,dx) := \sum_{k=1}^{n} \sum_{l=1}^{m} a_{k,l} q((t_{k-1}, t_k] \times A_{k,l}).$$

Note that the integral

$$\Sigma(F) \to L^1(F), \quad f \mapsto \iint_{\mathcal{X}} f(t,x)q(dt,dx) \tag{2.4.2}$$

is a linear operator.

Lemma 2.4.12 *For each $f \in \mathcal{L}^1_\beta(F)$ we have (2.4.3)*

$$\mathbb{E}\left[\iint_{\mathcal{X}} \|f(t,x)\|N(dt,dx)\right] = \iint_{\mathcal{X}} \|f(t,x)\|\beta(dx)dt \tag{2.4.3}$$

and the integral operator

$$\mathcal{L}^1_\beta(F) \to L^1(F), \quad f \mapsto \iint_{\mathcal{X}} f(t,x)q(dt,dx) \tag{2.4.4}$$

is continuous.

Proof We show that for any $g \in \mathcal{L}^1_\beta(\mathbb{R})$ with $g \geq 0$ we have

$$\mathbb{E}\left[\iint_{\mathcal{X}} g(t,x)N(dt,dx)\right] = \iint_{\mathcal{X}} g(t,x)\beta(dx)dt, \tag{2.4.5}$$

which yields (2.4.3). By inspection, (2.4.5) holds true for every simple function $g \in \Sigma(\mathbb{R})$ with $g \geq 0$. For an arbitrary $g \in \mathcal{L}^1_\beta(\mathbb{R})$ with $g \geq 0$ there exists a sequence $(g_n)_{n \in \mathbb{N}} \subset \Sigma(\mathbb{R})$ of simple functions such that $g_n \geq 0$ for all $n \in \mathbb{N}$ and $g_n \uparrow g$ everywhere. By the monotone convergence theorem we obtain

$$\mathbb{E}\left[\iint_{\mathcal{X}} g(t,x)N(dt,dx)\right] = \lim_{n \to \infty} \mathbb{E}\left[\iint_{\mathcal{X}} g_n(t,x)N(dt,dx)\right]$$

$$= \lim_{n \to \infty} \iint_{\mathcal{X}} g_n(t,x)\beta(dx)dt$$

$$= \iint_{\mathcal{X}} g(t,x)\beta(dx)dt,$$

proving (2.4.5). We deduce that for every $f \in \mathcal{L}^1_\beta(F)$ relation (2.4.3) is valid and

$$\mathbb{E}\left[\left\|\iint_{\mathcal{X}} f(t, x) N(dt, dx)\right\|\right] \leq \mathbb{E}\left[\iint_{\mathcal{X}} \|f(t, x)\| N(dt, dx)\right]$$

$$= \iint_{\mathcal{X}} \|f(t, x)\| \beta(dx) dt,$$

proving the continuity of the linear operator (2.4.4).   □

For $f \in \mathcal{L}_\beta^1(F)$ and $B \in \mathcal{E}$ we define

$$\int_0^t \int_B f(s, x) q(ds, dx) := \iint_{(0,t] \times B} f(s, x) q(ds, dx), \quad t \geq 0.$$

Lemma 2.4.13 *For each $f \in \mathcal{L}_\beta^1(F)$ and each $T \geq 0$ the process $M = (M_t)_{t \in [0,T]}$ given by*

$$M_t = \int_0^t \int_E f(t, x) q(dt, dx), \quad t \in [0, T]$$

belongs to $\mathcal{M}_T^1(F)$.

Proof There exists a sequence $(f_n)_{n \in \mathbb{N}} \subset \Sigma(F)$ of simple functions such that $f_n \to f$ in $\mathcal{L}_\beta^1(F)$. For $n \in \mathbb{N}$ we define $M^n = (M_t^n)_{t \in [0,T]}$ by

$$M_t^n = \int_0^t \int_E f_n(s, x) q(ds, dx), \quad t \in [0, T].$$

By Lemma 2.4.7 we have $M^n \in \mathcal{M}_T^1(F)$ for all $n \in \mathbb{N}$. By the continuity of the integral operator (2.4.4) we have $\mathbb{E}[\|M_T^n - M_T\|] \to 0$. Since $\mathcal{M}_T^1(F)$ is a Banach space according to Lemma 2.3.6, we deduce that $M \in \mathcal{M}_T^1(F)$.   □

Now we fix $T > 0$ and set

$$\mathcal{L}_\beta^1(\mathcal{P}_T; F) := L^1(\Omega \times [0, T] \times E, \mathcal{P}_T \otimes \mathcal{E}, \mathbb{P} \otimes \lambda \otimes \beta; F).$$

An analogous argument as in the proof of Lemma 2.4.12 provides the following result:

Lemma 2.4.14 *For each $f \in \mathcal{L}_\beta^1(\mathcal{P}_T; F)$ we have*

$$\mathbb{E}\left\|\iint_{\mathcal{X}} f(t, x) N(dt, dx)\right\| = \mathbb{E}\left[\iint_{\mathcal{X}} \|f(t, x)\| \beta(dx) dt\right] \qquad (2.4.6)$$

and the integral operator

$$\mathcal{L}_\beta^1(\mathcal{P}_T; F) \to L^1(F), \quad f \mapsto \iint_{\mathcal{X}} f(t, x) q(dt, dx) \qquad (2.4.7)$$

is continuous.

For $f \in \mathcal{L}_\beta^1(\mathcal{P}_T; F)$ and $B \in \mathcal{E}$ we define

$$\int_0^t \int_B f(s, x) q(ds, dx) := \iint_{(0,t] \times B} f(s, x) q(ds, dx), \quad t \in [0, T].$$

Lemma 2.4.15 *For each $f \in \mathcal{L}_\beta^1(\mathcal{P}_T; F)$ the process $M = (M_t)_{t \in [0,T]}$ given by*

$$M_t = \int_0^t \int_E f(t, x) q(dt, dx), \quad t \in [0, T]$$

belongs to $\mathcal{M}_T^1(F)$, and the integral operator

$$\mathcal{L}_\beta^1(\mathcal{P}_T; F) \to \mathcal{M}_T^1(F), \quad f \mapsto \left(\int_0^t \int_E f(s, x) q(ds, dx) \right)_{t \in [0,T]} \qquad (2.4.8)$$

is continuous.

Proof Arguing as in the proof of Lemma 2.4.13, we obtain, by using Lemma 2.4.8 instead of Lemma 2.4.7, that $M \in \mathcal{M}_T^1(F)$. The continuity of the integral operator (2.4.8) follows from Lemma 2.4.14. $\qquad\qquad\qquad\qquad\qquad\qquad\qquad\qquad\qquad\qquad\qquad \square$

2.5 Characteristic Functions

Let $(\Omega, \mathcal{F}, \mathbb{P})$ be a probability space and let E be a separable Banach space. For a random variable $X : \Omega \to E$ we define the *characteristic function*

$$\varphi_X : E^* \to \mathbb{C}, \quad \varphi_X(x^*) = \mathbb{E}[e^{i\langle x^*, X \rangle}].$$

Theorem 2.5.1 *Let $X, Y : \Omega \to E$ be two random variables with $\varphi_X = \varphi_Y$. Then we have $\mathbb{P}^X = \mathbb{P}^Y$.*

Proof We will show that $\mathbb{P}^X|_\mathcal{C} = \mathbb{P}^Y|_\mathcal{C}$, where $\mathcal{C}$ denotes the system of cylinder sets. Since $\mathcal{C}$ is stable under intersection and generates $\mathcal{B}(E)$ by Proposition 2.1.1, the uniqueness theorem for measures, see, e.g., [10, Theorem 5.4], then yields that $\mathbb{P}^X = \mathbb{P}^Y$.

Let $n \in \mathbb{N}$, the linear functionals $x_1^*, \ldots, x_n^* \in E^*$ and $u \in \mathbb{R}^n$ be arbitrary. Then we have

$$\varphi_{\langle x_k^*, X\rangle_{k=1,\ldots,n}}(u) = \mathbb{E}\left[\exp\left(i\sum_{k=1}^n u_k\langle x_k^*, X\rangle\right)\right] = \mathbb{E}\left[\exp\left(i\Big\langle\sum_{k=1}^n u_k x_k^*, X\Big\rangle\right)\right]$$

$$= \varphi_X\left(\sum_{k=1}^n u_k x_k^*\right) = \varphi_Y\left(\sum_{k=1}^n u_k x_k^*\right) = \mathbb{E}\left[\exp\left(i\Big\langle\sum_{k=1}^n u_k x_k^*, Y\Big\rangle\right)\right]$$

$$= \mathbb{E}\left[\exp\left(i\sum_{k=1}^n u_k\langle x_k^*, Y\rangle\right)\right] = \varphi_{\langle x_k^*, Y\rangle_{k=1,\ldots,n}}(u).$$

Using the uniqueness theorem for characteristic functions in finite dimensions, see, e.g., [47, Theorem 14.1], for all $B_1, \ldots, B_n \in \mathcal{B}(\mathbb{R})$ we obtain

$$\mathbb{P}^X(x_1^* \in B_1, \ldots, x_n^* \in B_n) = \mathbb{P}(\langle x_1^*, X\rangle \in B_1, \ldots, \langle x_n^*, X\rangle \in B_n)$$

$$= \mathbb{P}^{\langle x_k^*, X\rangle_{k=1,\ldots,n}}(B_1 \times \ldots \times B_n) = \mathbb{P}^{\langle x_k^*, Y\rangle_{k=1,\ldots,n}}(B_1 \times \ldots \times B_n)$$

$$= \mathbb{P}(\langle x_1^*, Y\rangle \in B_1, \ldots, \langle x_n^*, Y\rangle \in B_n) = \mathbb{P}^Y(x_1^* \in B_1, \ldots, x_n^* \in B_n),$$

showing that $\mathbb{P}^X = \mathbb{P}^Y$. $\square$

Hence, all well-known theorems for finite dimensional random variables transfer to Banach space valued random variables. In particular, we obtain the following result:

Theorem 2.5.2 Let $X_1, \ldots, X_n : \Omega \to F$ be random variables. Then $X_1, \ldots, X_n$ are independent if and only if for all $x_1^*, \ldots, x_n^* \in F^*$ we have

$$\varphi_{(X_1,\ldots,X_n)}(x_1^*, \ldots, x_n^*) = \prod_{j=1}^n \varphi_{X_j}(x_j^*).$$

Now let N be a Poisson random measure.

Proposition 2.5.3 Let F be a separable Banach space. For each $f \in \mathcal{L}_\beta^1(F)$ the characteristic function of the integral

$$N_f = \iint_{\mathcal{X}} f(s, x) N(ds, dx)$$

is given by

$$\mathbb{E}[e^{i\langle y^*, N_f\rangle}] = \exp\left(\iint_{\mathcal{X}}\left(e^{i\langle y^*, f(s,x)\rangle} - 1\right)\beta(dx)ds\right), \quad y^* \in E^*. \qquad (2.5.1)$$

Proof Let $f \in \Sigma(F)$ be an arbitrary simple function of the form (2.4.1). By Theorem 2.4.6 the random variables $N((t_{k-1}, t_k] \times A_{k,l})$ are independent and have a Poisson distribution with mean $\lambda((t_{k-1}, t_k])\beta(A_{k,l})$. Thus, for every $y^* \in F^*$ we obtain

$$
\begin{aligned}
\mathbb{E}[e^{i\langle y^*, N_f \rangle}] &= \mathbb{E}\left[\exp\left(\left\langle iy^*, \sum_{k=1}^{n} \sum_{l=1}^{m} a_{k,l} N((t_{k-1}, t_k] \times A_{k,l}) \right\rangle \right) \right] \\
&= \prod_{k=1}^{n} \prod_{l=1}^{m} \mathbb{E}\left[\exp\left(i\langle y^*, a_{k,l} \rangle N((t_{k-1}, t_k] \times A_{k,l}) \right) \right] \\
&= \prod_{k=1}^{n} \prod_{l=1}^{m} \exp\left(\lambda((t_{k-1}, t_k])\beta(A_{k,l})\left(e^{i\langle y^*, a_{k,l} \rangle} - 1 \right) \right) \\
&= \exp\left(\sum_{k=1}^{n} \sum_{l=1}^{m} \left(e^{i\langle y^*, a_{k,l} \rangle} - 1 \right) \lambda((t_{k-1}, t_k])\beta(A_{k,l}) \right) \\
&= \exp\left(\iint_{\mathcal{X}} \left(e^{i\langle y^*, f(s,x) \rangle} - 1 \right) \beta(dx) ds \right).
\end{aligned}
$$

This proves (2.5.1) for every $f \in \Sigma(F)$. Now, let $f \in \mathcal{L}_{\beta}^1(F)$ be arbitrary. There exists a sequence $(f_n)_{n \in \mathbb{N}} \subset \Sigma(F)$ of simple functions such that $f_n \to f$ in $\mathcal{L}_{\beta}^1(F)$. By the continuity of the integral operator (2.4.4) we have $N_{f_n} \to N_f$ in $L^1(\Omega, \mathcal{F}, \mathbb{P}; F)$. Let $y^* \in F^*$ be arbitrary. There exists a subsequence $(n_k)_{k \in \mathbb{N}}$ such that $\langle y^*, f_{n_k} \rangle \to \langle y^*, f \rangle$ almost surely and $\langle y^*, N_{f_{n_k}} \rangle \to \langle y^*, N_f \rangle$ almost surely. By Lebesgue's dominated convergence theorem we have

$$
\mathbb{E}[e^{i\langle y^*, N_{f_{n_k}} \rangle}] \to \mathbb{E}[e^{i\langle y^*, N_f \rangle}].
$$

Note that for all $x \in \mathbb{R}$ we have

$$
|e^{ix} - 1| = |\cos x - 1 + i \sin x| = \sqrt{(\cos x - 1)^2 + \sin^2 x}
$$

$$
\leq \sqrt{x^2 + x^2} = \sqrt{2}|x|.
$$

Therefore, for every $g \in \mathcal{L}_{\beta}^1(E)$ we have

$$
|e^{i\langle y^*, g(s,x) \rangle} - 1| \leq \sqrt{2}\|y^*\| \|g(s,x)\|, \quad (s,x) \in \mathcal{X}.
$$

Using the generalized Lebesgue dominated convergence theorem (Lemma 7.1.8) we deduce

$$
\iint_{\mathcal{X}} \left(e^{i\langle y^*, f_{n_k}(s,x) \rangle} - 1 \right) \beta(dx) ds \to \iint_{\mathcal{X}} \left(e^{i\langle y^*, f(s,x) \rangle} - 1 \right) \beta(dx) ds.
$$

Consequently, the identity (2.5.1) is valid for all $f \in \mathcal{L}_{\beta}^1(F)$. $\square$

Lemma 2.5.4 *Let F, G be two separable Banach spaces. Let $f \in \mathcal{L}_\beta^1(F)$, $g \in \mathcal{L}_\beta^1(G)$ and $A, B \in \mathcal{B}(\mathcal{X})$ with $A \cap B = \emptyset$ be arbitrary. Then, the random variables*

$$X = \iint_A f(t, x) N(dt, dx) \quad and \quad Y = \iint_B g(t, x) N(dt, dx)$$

are independent.

Proof First we assume that $f \in \Sigma(F)$ and $g \in \Sigma(G)$ are simple functions of the form

$$f(t, x) = \sum_{k=1}^{n} \sum_{l=1}^{m} a_{k,l} \mathbb{1}_{A_{k,l}}(x) \mathbb{1}_{(t_{k-1}, t_k]}(t), \tag{2.5.2}$$

$$g(t, x) = \sum_{k=1}^{p} \sum_{l=1}^{q} B_{k,l} \mathbb{1}_{B_{k,l}}(x) \mathbb{1}_{(s_{k-1}, s_k]}(t) \tag{2.5.3}$$

with disjoint sets $A_{k,l} \subset A$ and $B_{k,l} \subset B$. Then we have

$$X = \sum_{k=1}^{n} \sum_{l=1}^{m} a_{k,l} N((t_{k-1}, t_k] \times A_{k,l}) \quad and \quad Y = \sum_{k=1}^{p} \sum_{l=1}^{q} b_{k,l} N((s_{k-1}, s_k] \times B_{k,l}).$$

Using Theorem 2.4.6, the random variables X and Y are independent. In the general case where $f \in \mathcal{L}_\beta^1(F)$ and $g \in \mathcal{L}_\beta^1(G)$ there exist sequences $(f_n)_{n \in \mathbb{N}} \subset \Sigma(F)$ and $(g_n)_{n \in \mathbb{N}} \subset \Sigma(G)$ of simple functions of the type (2.5.2) and (2.5.3) such that $f_n \to f$ in $\mathcal{L}_\beta^2(F)$ and $g_n \to g$ in $\mathcal{L}_\beta^2(G)$. Then, for each $n \in \mathbb{N}$ the random variables

$$X_n = \iint_A f_n(t, x) N(dt, dx) \quad and \quad Y_n = \iint_B g_n(t, x) N(dt, dx)$$

are independent. By the continuity of the integral operator (2.4.4) we have $X_n \to X$ in $\mathcal{L}_\beta^1(F)$ and $Y_n \to Y$ in $\mathcal{L}_\beta^1(G)$. Consequently, the random variables X and Y are also independent. $\square$

2.6 Remarks and Related Literature

The material of Sect. 2.1 is from [23]. In Sect. 2.4, we have taken the general definition of a Poisson random measure, which can be found in [48]. The inequalities in Sect. 2.3 are due to Doob and follow from the fact that $\|M_t\|$ is a submartingale.

Chapter 3
Stochastic Integrals with Respect to Compensated Poisson Random Measures

In this chapter, we define the stochastic integral. Throughout this section, $(\Omega, \mathcal{F}, (\mathcal{F}_t)_{t\geq 0}, \mathbb{P})$ is a filtered probability space.

3.1 The Wiener Integral with Respect to Compensated Poisson Random Measures

Let $(E, \mathcal{E})$ be a Blackwell space and let $q(dt, dx)$ be a compensated Poisson random measure on $\mathcal{X} = \mathbb{R}_+ \times E$ with compensator $\nu(dt, dx) = dt \otimes \beta(dx)$. Let F be a separable Banach space.

Let $\mathcal{L}^2_\beta(F) = L^2(\mathcal{X}, \mathcal{B}(\mathcal{X}), \lambda \otimes \beta; F)$. As discussed in Sect. 2.4 the σ-algebra $\mathcal{B}(\mathcal{X})$ is generated by the semiring

$$\mathcal{S} = \{\{0\} \times B : B \in \mathcal{E}\} \cup \{(s, t] \times B : 0 \leq s < t \text{ and } B \in \mathcal{E}\}$$

and by Proposition 2.1.6 the linear space $\Sigma(F) = \Sigma(\mathcal{S}; F)$ is dense in $\mathcal{L}^2_\beta(F)$. Moreover, any function $f \in \Sigma(F)$ is of the form (2.4.1).

For every $f \in \Sigma(F)$ of the form (2.4.1) we defined in Definition 2.4.11 the *Wiener integral* of f with respect to q as

$$\iint_{\mathcal{X}} f(t, x) q(dt, dx) := \sum_{k=1}^{n} \sum_{l=1}^{m} a_{k,l} q((t_{k-1}, t_k] \times A_{k,l}).$$

Note that the Wiener integral

$$\Sigma(F) \rightarrow L^2(F), \quad f \mapsto \iint_{\mathcal{X}} f(t, x) q(dt, dx) \tag{3.1.1}$$

is a linear operator.

© Springer International Publishing Switzerland 2015
V. Mandrekar and B. Rüdiger, *Stochastic Integration in Banach Spaces*,
Probability Theory and Stochastic Modelling 73, DOI 10.1007/978-3-319-12853-5_3

Remark 3.1.1 If $F = H$ is a separable Hilbert space, then for simple functions we have the so-called Itô isometry

$$\mathbb{E}\left[\left\|\iint_{\mathcal{X}} f(t, x)q(dt, dx)\right\|^2\right] = \iint_{\mathcal{X}} \|f(t, x)\|^2 \beta(dx)dt \quad for\,all\,f \in \Sigma(H). \quad (3.1.2)$$

Indeed, for a simple $f \in \Sigma(H)$ of the form (2.4.1) we have

$$\mathbb{E}\left[\left\|\iint_{\mathcal{X}} f(t, x)q(dt, dx)\right\|^2\right]$$

$$= \mathbb{E}\left[\left\|\sum_{k=1}^{n}\sum_{l=1}^{m} a_{k,l}q((t_{k-1}, t_k] \times A_{k,l})\right\|^2\right]$$

$$= \mathbb{E}\left[\left\langle \sum_{k=1}^{n}\sum_{l=1}^{m} a_{k,l}q((t_{k-1}, t_k] \times A_{k,l}), \sum_{k=1}^{n}\sum_{l=1}^{m} a_{k,l}q((t_{k-1}, t_k] \times A_{k,l})\right\rangle\right]$$

$$= \sum_{k=1}^{n}\sum_{l=1}^{m} \|a_{k,l}\|^2 \mathbb{E}[q((t_{k-1}, t_k] \times A_{k,l})]$$

$$+ \sum_{k=1}^{n}\sum_{l=1}^{m}\sum_{i=1}^{n}\sum_{j=1}^{m} a_{k,l}a_{i,j}\mathbb{E}[\langle q((t_{k-1}, t_k] \times A_{k,l}), q((t_{i-1}, t_i] \times A_{i,j})\rangle].$$

Using Theorem 2.4.6 we obtain

$$\sum_{k=1}^{n}\sum_{l=1}^{m} \|a_{k,l}\|^2 \mathbb{E}[q((t_{k-1}, t_k] \times A_{k,l})] = \sum_{k=1}^{n}\sum_{l=1}^{m} \|a_{k,l}\|^2 \beta(A_{k,l})\lambda((t_{k-1}, t_k])$$

$$= \iint_{\mathcal{X}} \|f(t, x)\|^2 \beta(dx)dt.$$

For $k < i$ the random variable $q((t_{i-1}, t_i] \times A_{i,j})$ is independent of $\mathcal{F}_{t_{i-1}}$ and $q((t_{k-1}, t_k] \times A_{k,l})$ is $\mathcal{F}_{t_{i-1}}$-measurable. Therefore, we get

$$\mathbb{E}[\langle q((t_{k-1}, t_k] \times A_{k,l}), q((t_{i-1}, t_i] \times A_{i,j})\rangle]$$

$$= \mathbb{E}[\mathbb{E}[\langle q((t_{k-1}, t_k] \times A_{k,l}), q((t_{i-1}, t_i] \times A_{i,j})\rangle] \mid \mathcal{F}_{t_{i-1}}]$$

$$= \mathbb{E}[\langle q((t_{k-1}, t_k] \times A_{k,l}), \mathbb{E}[q((t_{i-1}, t_i] \times A_{i,j})]\rangle] = 0,$$

and hence, the Itô isometry (3.1.2) is valid.

Consequently, if $F = H$ is a separable Hilbert space, then the integral operator (3.1.1) is an isometry, and therefore in particular continuous. Thus, and because $\Sigma(H)$ is dense in $\mathcal{L}^2_\beta(H)$, it has a unique extension

$$\mathcal{L}^2_\beta(H) \to L^2(H), \quad f \mapsto \iint\limits_{\mathcal{X}} f(t, x) q(dt, dx),$$

which we also call the *Wiener integral*, and we have the Itô isometry

$$\mathbb{E}\left[\left\|\iint\limits_{\mathcal{X}} f(t, x) q(dt, dx)\right\|^2\right] = \iint\limits_{\mathcal{X}} \|f(t, x)\|^2 \beta(dx) dt \quad \text{for all} f \in \mathcal{L}^2_\beta(H). \quad (3.1.3)$$

If F is a general Banach space, then the definition of the Wiener integral becomes more involved, because relation (3.1.2) may not be satisfied for all $f \in \Sigma(F)$. The article [102] gives a counterexample for stochastic integrals with respect to a one-dimensional Brownian motion.

However, if there exists a constant $K_\beta > 0$ (which may depend on β) such that

$$\mathbb{E}\left[\left\|\iint\limits_{\mathcal{X}} f(t, x) q(dt, dx)\right\|^2\right] \leq K_\beta \iint\limits_{\mathcal{X}} \|f(t, x)\|^2 \beta(dx) dt \quad \text{for all} f \in \Sigma(F),$$

$$(3.1.4)$$

then we can analogously define the Wiener integral for all $f \in \mathcal{L}^2_\beta(F)$ as the continuous linear operator

$$\mathcal{L}^2_\beta(F) \to L^2(\Omega, \mathcal{F}, \mathbb{P}; F), \quad f \mapsto \iint\limits_{\mathcal{X}} f(t, x) q(dt, dx), \quad (3.1.5)$$

which is the unique extension of (3.1.1). In particular, we obtain the estimate

$$\mathbb{E}\left[\left\|\iint\limits_{\mathcal{X}} f(t, x) q(dt, dx)\right\|^2\right] \leq K_\beta \iint\limits_{\mathcal{X}} \|f(t, x)\|^2 \beta(dx) dt \quad \text{for all} f \in \mathcal{L}^2_\beta(F).$$

Remark 3.1.2 The Wiener Integral in (3.1.5) is cádlág.

We proceed with the definition of the Pettis integral. Let $f : \mathcal{X} \to F$ be a measurable function such that $\langle y^*, f \rangle \in \mathcal{L}^2_\beta(\mathbb{R})$ for all $y^* \in F^*$. We define the linear operator

$$T^f : F^* \to L^2(\Omega, \mathcal{F}, \mathbb{P}), \quad T^f y^* := \iint\limits_{\mathcal{X}} \langle y^*, f(s, x) \rangle q(ds, dx).$$

Arguing as in the proof of Dunford's lemma (see Lemma 2.1.7), we show that T^f is continuous. The function f is said to be *Pettis integrable* if there exists a random variable $Z^f \in L^2(\Omega, \mathcal{F}, \mathbb{P}; F)$ such that almost surely

$$T^f y^* = \langle y^*, Z^f \rangle. \tag{3.1.6}$$

Note that such a random variable Z^f is $\mathbb{P}$-almost surely unique, provided it exists, and that the set of Pettis integrable functions forms a linear space. Following the ideas of [90], we call Z^f the *Pettis integral* of f and set

$$(\text{P}-) \iint\limits_{\mathcal{X}} f(s, x) q(ds, dx) := Z^f.$$

We observe that for each simple function $f \in \Sigma(F)$ the Pettis integral exists and coincides with the Wiener integral, that is

$$\iint\limits_{\mathcal{X}} f(s, x) q(ds, dx) = (\text{P}-) \iint\limits_{\mathcal{X}} f(s, x) q(ds, dx) \quad \text{for all } f \in \Sigma(F). \tag{3.1.7}$$

Lemma 3.1.3 *Suppose there exists a constant $K_\beta > 0$ such that (3.1.4) is satisfied. Then each function $f \in \mathcal{L}^2_\beta(F)$ is Pettis integrable and we have*

$$\iint\limits_{\mathcal{X}} f(s, x) q(ds, dx) = (\text{P}-) \iint\limits_{\mathcal{X}} f(s, x) q(ds, dx) \quad \text{for all } f \in \mathcal{L}^2_\beta(F). \tag{3.1.8}$$

Moreover, for each $x^* \in F^*$ we have

$$\left\langle x^*, \iint\limits_{\mathcal{X}} f(s, x) q(ds, dx) \right\rangle = \iint\limits_{\mathcal{X}} \langle x^*, f(s, x) \rangle q(ds, dx) \quad \text{for all } f \in \mathcal{L}^2_\beta(F). \tag{3.1.9}$$

Proof The identity (3.1.9) is immediately verified for simple functions $f \in \Sigma(F)$, and thus follows for general functions $f \in \mathcal{L}^2_\beta(F)$ by choosing an approximating sequence $(f_n)_{n \in \mathbb{N}} \subset \Sigma(F)$ and passing to the limit. Because of (3.1.9), for an arbitrary function $f \in \mathcal{L}^2_\beta(F)$ the random variable

$$Z^f := \iint\limits_{\mathcal{X}} f(s, x) q(ds, dx) \tag{3.1.10}$$

satisfies (3.1.6), proving the identity (3.1.8). □

Now the question arises which properties of the continuous linear operator T^f ensure that the function $f : \mathcal{X} \to F$ is Pettis integrable. If F is a Hilbert space, we get a straight answer. Recall that for two separable Hilbert spaces H_1 and H_2 a linear operator $T : H_1 \rightarrow H_2$ is called a Hilbert–Schmidt operator if for one (and thus for any) orthonormal basis $(e_j)_{j \in \mathbb{N}}$ of H_1 we have

$$\sum_{j \in \mathbb{N}} \|Te_j\|^2 < \infty.$$

Theorem 3.1.4 *Suppose that $F = H$ is a separable Hilbert space, and let $f : \mathcal{X} \rightarrow H$ be a measurable function with $\langle y, f \rangle \in \mathcal{L}_\beta^2(\mathbb{R})$ for all $y \in H$. Then the following statements are equivalent:*

1. *$T^f : H \rightarrow L^2(\Omega, \mathcal{F}, \mathbb{P})$ is a Hilbert–Schmidt operator.*
2. *$f \in \mathcal{L}_\beta^2(H)$.*
3. *The function f is Pettis integrable with*

$$\mathbb{E}\left[\left\| (P-) \iint_{\mathcal{X}} f(s, x) q(ds, dx) \right\|^2 \right] < \infty. \tag{3.1.11}$$

Proof Let $(e_j)_{j \in \mathbb{N}}$ be an orthonormal basis of H. By the Itô isometry, the monotone convergence theorem and Parseval's identity we have

$$\sum_{j \in \mathbb{N}} \|T^f e_j\|_{L^2}^2 = \sum_{j \in \mathbb{N}} \mathbb{E}[|T^f e_j|^2] = \sum_{j \in \mathbb{N}} \mathbb{E}\left[\left| \iint_{\mathcal{X}} \langle e_j, f(s, x) \rangle q(ds, dx) \right|^2 \right]$$

$$= \sum_{j \in \mathbb{N}} \iint_{\mathcal{X}} |\langle e_j, f(s, x) \rangle|^2 \beta(dx) ds = \iint_{\mathcal{X}} \sum_{j \in \mathbb{N}} |\langle e_j, f(s, x) \rangle|^2 \beta(dx) ds$$

$$= \iint_{\mathcal{X}} \|f(s, x)\|^2 \beta(dx) ds,$$

proving (1) $\Leftrightarrow$ (2). The implication (2) $\Rightarrow$ (3) follows from Lemma 3.1.3. Suppose that f is Pettis integrable with relation (3.1.11) being satisfied. By the monotone convergence theorem and Parseval's identity we obtain

$$\sum_{j \in \mathbb{N}} \|T^f e_j\|_{L^2}^2 = \sum_{j \in \mathbb{N}} \mathbb{E}[|T^f e_j|^2] = \mathbb{E}\left[\sum_{j \in \mathbb{N}} |T^f e_j|^2 \right]$$

$$= \mathbb{E}\left[\sum_{j \in \mathbb{N}} |\langle e_j, Z^f \rangle|^2 \right]$$

$$= \mathbb{E}[\|Z^f\|^2] = \mathbb{E}\left[\left\| (P-) \iint_{\mathcal{X}} f(s, x) q(ds, dx) \right\|^2 \right] < \infty,$$

providing (3) $\Rightarrow$ (1). $\qquad \square$

Theorem 3.1.5 *Let F be a separable Banach space. The following statements are equivalent:*

1. *Each $f \in \mathcal{L}_\beta^2(F)$ is Pettis integrable and we have* (3.1.11).
2. *Each $f \in \mathcal{L}_\beta^2(F)$ is Pettis integrable, we have* (3.1.11) *and the linear operator*

$$\mathcal{L}_\beta^2(F) \to L^2(\Omega, \mathcal{F}, \mathbb{P}; F), \quad f \mapsto (\text{P}-)\iint_{\mathcal{X}} f(s, x)q(ds, dx) \qquad (3.1.12)$$

is continuous.

3. *There exists a constant $K_\beta > 0$ such that* (3.1.4) *is satisfied.*

If the previous conditions are fulfilled, then we have identities (3.1.8) *and* (3.1.9).

Proof The implication $(2) \Rightarrow (1)$ is obvious and the implication $(3) \Rightarrow (1)$ as well as the additional statement follow from Lemma 3.1.3. The implication $(2) \Rightarrow (3)$ is valid, because the Wiener integral and the Pettis integral coincide for simple functions $f \in \Sigma(F)$, see (3.1.7).

Consequently, it only remains to prove $(1) \Rightarrow (2)$. We shall prove that the linear operator

$$S : \mathcal{L}_\beta^2(F) \to L^2(\Omega, \mathcal{F}, \mathbb{P}; F), \quad f \mapsto Z^f,$$

is a closed operator. Then the assertion follows from the closed graph theorem. Let $(f_n)_{n \in \mathbb{N}} \subset \mathcal{L}_\beta^2(F)$ and $f \in \mathcal{L}_\beta^2(F)$ be such that $f_n \to f$ and $Sf_n \to g$ for some $g \in L^2(\Omega, \mathcal{F}, \mathbb{P}; F)$. Then, for each $y^* \in F^*$ we have

$$\lim_{n \to \infty} \langle y^*, Sf_n \rangle = \langle y^*, g \rangle$$

and by (3.1.6) and the continuity of the linear operator

$$\mathcal{L}_\beta^2(\mathbb{R}) \to L^2(\Omega, \mathcal{F}, \mathbb{P}; \mathbb{R}), \quad h \mapsto \iint_{\mathcal{X}} h(t, x)q(dt, dx),$$

we obtain almost surely

$$\lim_{n \to \infty} \langle y^*, Sf_n \rangle = \lim_{n \to \infty} \langle y^*, Z^{f_n} \rangle = \lim_{n \to \infty} T^{f_n} y^*$$

$$= \lim_{n \to \infty} \iint_{\mathcal{X}} \langle y^*, f_n(t, x) \rangle q(dt, dx) = \iint_{\mathcal{X}} \langle y^*, f(t, x) \rangle q(dt, dx)$$

$$= T^f y^* = \langle y^*, Z^f \rangle = \langle y^*, Sf \rangle.$$

We deduce that $Sf = g$, proving that S is a closed operator. $\qquad \square$

Let us now introduce Banach spaces of type 2, which are more general than Hilbert spaces.

Definition 3.1.6 A separable Banach space F is called a *Banach space of type 2* if there exists a constant $K > 0$ such that for every probability space $(\Omega, \mathcal{F}, \mathbb{P})$, for each $n \in \mathbb{N}$ and for any collection $X_1, \ldots, X_n : \Omega \to F$ of independent, symmetric, Bochner-integrable random variables with $\mathbb{E}[X_i] = 0$, $i = 1, \ldots, n$, we have

$$\mathbb{E}\left[\left\|\sum_{i=1}^{n} X_i\right\|^2\right] \le K \sum_{i=1}^{n} \mathbb{E}[\|X_i\|^2]. \tag{3.1.13}$$

Remark 3.1.7 Note that every separable Hilbert space H is a Banach space of type 2. Indeed, for any independent random variables $X_1, \ldots, X_n : \Omega \to F$ with $\mathbb{E}[X_i] = 0$, $i = 1, \ldots, n$, we have

$$\mathbb{E}\left[\left\|\sum_{i=1}^{n} X_i\right\|^2\right] = \mathbb{E}\left[\left\langle \sum_{i=1}^{n} X_i, \sum_{i=1}^{n} X_i \right\rangle\right] = \sum_{i=1}^{n}\sum_{j=1}^{n} \mathbb{E}[\langle X_i, X_j \rangle]$$

$$= \sum_{i=1}^{n} \mathbb{E}[\|X_i\|^2] + 2 \sum_{i<j} \mathbb{E}[\langle X_i, X_j \rangle].$$

Since for $i < j$ we have, by the independence of X_i and X_j,

$$\mathbb{E}[\langle X_i, X_j \rangle] = \mathbb{E}[\mathbb{E}[\langle X_i, X_j \rangle \mid X_i]] = \mathbb{E}[\mathbb{E}[\langle x, X_j \rangle]|_{x=X_i}]$$
$$= \mathbb{E}[\langle x, \mathbb{E}[X_j] \rangle|_{x=X_i}] = 0,$$

we arrive at

$$\mathbb{E}\left[\left\|\sum_{i=1}^{n} X_i\right\|^2\right] = \sum_{i=1}^{n} \mathbb{E}[\|X_i\|^2],$$

showing that (3.1.13) is fulfilled.

It is easy to show that the Lebesgue space L^p on $\mathbb{R}$ for $2 \le p < \infty$ is of type 2 using Kahane inequality.

From now on, let F be a Banach space of type 2. Let $f \in \Sigma(F)$ be an arbitrary simple function of the form (2.4.1). Since the product sets $A_{k,l} \times (t_{k-1}, t_k]$ are disjoint, using Theorem 2.4.6 we obtain

$$\mathbb{E}\left[\left\|\iint_{\mathcal{X}} f(t, x) q(dt, dx)\right\|^2\right] = \mathbb{E}\left[\left\|\sum_{k=1}^{n}\sum_{l=1}^{m} a_{k,l} q((t_{k-1}, t_k] \times A_{k,l})\right\|^2\right]$$

$$\le K \sum_{k=1}^{n}\sum_{l=1}^{m} \|a_{k,l}\|^2 \mathbb{E}[q((t_{k-1}, t_k] \times A_{k,l})^2]$$

$$= K \sum_{k=1}^{n}\sum_{l=1}^{m} \|a_{k,l}\|^2 \beta(A_{k,l}) \lambda((t_{k-1}, t_k])$$

$$= K \iint_{\mathcal{X}} \|f(t,x)\|^2 \beta(dx)dt.$$

This shows that estimate (3.1.4) is satisfied with a constant not depending on β. Hence, we can define the Wiener integral as the continuous linear operator

$$\mathcal{L}_\beta^2(F) \to L^2(\Omega, \mathcal{F}, \mathbb{P}; F), \quad f \mapsto \iint_{\mathcal{X}} f(t,x)q(dt,dx), \qquad (3.1.14)$$

which is the unique extension of (3.1.1). In particular, we get

$$\mathbb{E}\left[\left\|\iint_{\mathcal{X}} f(t,x)q(dt,dx)\right\|^2\right] \le K \iint_{\mathcal{X}} \|f(t,x)\|^2 dt\beta(dx) \quad \text{for all} f \in \mathcal{L}_\beta^2(F).$$

$$(3.1.15)$$

For $f \in \mathcal{L}_\beta^2(F)$ and $A \in \mathcal{B}(\mathcal{X})$ we define

$$\iint_A f(t,x)q(dt,dx) := \iint_{\mathcal{X}} f(t,x)\mathbb{1}_A(t,x)q(dt,dx),$$

and for $B \in \mathcal{E}$ and $t \ge 0$ we define

$$\int_0^t \int_B f(s,x)q(ds,dx) := \iint_{(0,t]\times B} f(s,x)q(ds,dx).$$

Similar to the proof of Lemma 2.5.4, we obtain the following result:

Lemma 3.1.8 *Let F, G be two Banach spaces of type 2. Let $f \in \mathcal{L}_\beta^2(F)$, $g \in \mathcal{L}_\beta^2(G)$ and $A, B \in \mathcal{B}(X)$ with $A \cap B = \emptyset$ be arbitrary. Then the random variables*

$$X = \iint_A f(dt,dx)q(dt,dx) \quad and \quad Y = \iint_B g(dt,dx)q(dt,dx)$$

are independent.

Proposition 3.1.9 *Let E be a Banach space of type 2. For each $f \in \mathcal{L}_\beta^2(F)$ the characteristic function of the Wiener integral*

$$W_f = \iint_{\mathcal{X}} f(s,x)q(ds,dx)$$

is given by

$$\mathbb{E}[e^{i\langle y^*, W_f\rangle}] = \exp\left(\iint\limits_{\mathcal{X}} \left(e^{i\langle y^* f(s,x)\rangle} - 1 - i\langle y^*, f(s,x)\rangle\right)\beta(dx)ds\right), \quad y^* \in E^*.$$

(3.1.16)

Proof Let $f \in \Sigma(F)$ be an arbitrary simple function of the form (2.4.1). By Theorem 2.4.6 the random variables $N((t_{k-1}, t_k] \times A_{k,l})$ are independent and have a Poisson distribution with mean $\lambda((t_{k-1}, t_k])\beta(A_{k,l})$. Thus, for every $y^* \in F^*$ we obtain

$$\mathbb{E}[e^{i\langle y^*, W_f\rangle}] = \mathbb{E}\left[\exp\left(\left\langle i, \sum_{k=1}^{n}\sum_{l=1}^{m} a_{k,l}q((t_{k-1}, t_k] \times A_{k,l})\right\rangle\right)\right]$$

$$= \prod_{k=1}^{n}\prod_{l=1}^{m}\mathbb{E}\left[\exp\left(i\langle y^*, a_{k,l}\rangle q((t_{k-1}, t_k] \times A_{k,l})\right)\right]$$

$$= \prod_{k=1}^{n}\prod_{l=1}^{m}\exp\left(\lambda((t_{k-1}, t_k])\beta(A_{k,l})\left(e^{i\langle y^*, a_{k,l}\rangle} - 1\right)\right)$$

$$\times \exp\left(-i\langle y^*, a_{k,l}\rangle\lambda((t_{k-1}, t_k])\beta(A_{k,l})\right)$$

$$= \exp\left(\sum_{k=1}^{n}\sum_{l=1}^{m}\left(e^{i\langle y^*, a_{k,l}\rangle} - 1 - i\langle y^*, a_{k,l}\rangle\right)\lambda((t_{k-1}, t_k])\beta(A_{k,l})\right)$$

$$= \exp\left(\iint\limits_{\mathcal{X}}\left(e^{i\langle y^* f(s,x)\rangle} - 1 - i\langle y^*, f(s,x)\rangle\right)\beta(dx)ds\right).$$

This proves (3.1.16) for every $f \in \Sigma(F)$. Now, let $f \in \mathcal{L}_\beta^2(F)$ be arbitrary. There exists a sequence $(f_n)_{n\in\mathbb{N}} \subset \Sigma(F)$ of simple functions such that $f_n \to f$ in $\mathcal{L}_\beta^2(F)$. By the continuity of the Itô integral (3.1.14) we have $W_{f_n} \to W_f$ in $L^2(\Omega, \mathcal{F}, \mathbb{P}; F)$. Let $y^* \in F^*$ be arbitrary. There exists a subsequence $(n_k)_{k\in\mathbb{N}}$ such that $\langle y^*, f_{n_k}\rangle \to \langle y^*, f\rangle$ almost surely and $\langle y^*, W_{f_{n_k}}\rangle \to \langle y^*, W_f\rangle$ almost surely. By Lebesgue's dominated convergence theorem we have

$$\mathbb{E}[e^{i\langle y^*, W_{f_{n_k}}\rangle}] \to \mathbb{E}[e^{i\langle y^*, W_f\rangle}].$$

Note that for all $x \in \mathbb{R}$ we have

$$|e^{ix} - 1 - ix| = |\cos x - 1 + i(\sin x - x)| = \sqrt{(\cos x - 1)^2 + (\sin x - x)^2}$$

$$\leq \sqrt{\left(\frac{x^2}{2}\right)^2 + \left(\frac{x^2}{2}\right)^2} \leq \frac{1}{\sqrt{2}}x^2.$$

Therefore, for every $g \in \mathcal{L}^2_\beta(E)$ we have

$$|e^{i\langle y^*, g(s,x)\rangle} - 1 - i\langle y^*, g(s,x)\rangle| \leq \frac{1}{\sqrt{2}}\|y^*\|^2 \|g(s,x)\|^2, \quad (s,x) \in \mathcal{X}.$$

Using the generalized Lebesgue dominated convergence theorem (Lemma 7.1.8) we deduce

$$\iint\limits_{\mathcal{X}} \left(e^{i\langle y^*, f_{n_k}(s,x)\rangle} - 1 - i\langle y^*, f_{n_k}(s,x)\rangle \right) \beta(dx)ds$$

$$\to \iint\limits_{\mathcal{X}} \left(e^{i\langle y^*, f(s,x)\rangle} - 1 - i\langle y^*, f(s,x)\rangle \right) \beta(dx)ds.$$

Consequently, the identity (3.1.16) is valid for all $f \in \mathcal{L}^2_\beta(F)$.   □

Lemma 3.1.10 *For each $f \in \mathcal{L}^2_\beta(F)$ and each $T > 0$ the process $M = (M_t)_{t \in [0,T]}$ given by*

$$M_t = \int\limits_0^t \int\limits_E f(s,x)q(ds,dx), \quad t \in [0,T]$$

belongs to $\mathcal{M}^2_T(F)$.

Proof There exists a sequence $(f_n)_{n \in \mathbb{N}} \subset \Sigma(F)$ of simple functions such that $f_n \to f$ in $\mathcal{L}^2_\beta(F)$. For $n \in \mathbb{N}$ we define $M^n = (M^n_t)_{t \in [0,T]}$ by

$$M^n_t = \int\limits_0^t \int\limits_E f_n(s,x)q(ds,dx), \quad t \in [0,T].$$

By Lemma 2.4.7 we have $M^n \in \mathcal{M}^2_T(F)$ for all $n \in \mathbb{N}$. By the continuity of the integral operator (3.1.14) we have $\mathbb{E}[\|M^n_T - M_T\|^2] \to 0$. Since $\mathcal{M}^2_T(F)$ is a Banach space according to Lemma 2.3.6, we deduce that $M \in \mathcal{M}^2_T(F)$.   □

For the rest of this section, let F be separable Banach space. The integrals in the upcoming lemma are Bochner integrals.

Lemma 3.1.11 *Suppose there is a constant $K_\beta > 0$ such that (3.1.4) is satisfied. Then for each $f \in \mathcal{L}^1_\beta(F) \cap \mathcal{L}^2_\beta(F)$ we have*

$$\iint\limits_{\mathcal{X}} f(t,x)q(dt,dx) = \iint\limits_{\mathcal{X}} f(t,x)N(dt,dx) - \iint\limits_{\mathcal{X}} f(t,x)\beta(dx)dt. \qquad (3.1.17)$$

Proof For every simple function $f \in \Sigma(F)$ identity (3.1.17) holds true by inspection. By Proposition 2.1.6, the linear space $\Sigma(F)$ is dense in $\mathcal{L}_\beta^1(F)$ and in $\mathcal{L}_\beta^2(F)$. The continuity of the integral operators (3.1.14) and (2.4.4) yields that (3.1.17) is valid for all $f \in \mathcal{L}_\beta^1(F) \cap \mathcal{L}_\beta^2(F)$.   □

Lemma 3.1.12 *For all $f \in \mathcal{L}_\beta^1(F) \cap \mathcal{L}_\beta^2(F)$ we have*

$$\mathbb{E}\left[\left\|\iint_{\mathcal{X}} f(s,x)q(ds,dx)\right\|^2\right]$$

$$\leq 4\iint_{\mathcal{X}} \|f(s,x)\|^2 \beta(dx)ds + 6\left(\iint_{\mathcal{X}} \|f(s,x)\|\beta(dx)ds\right)^2.$$

Proof Using Lemma 3.1.11 we obtain

$$\mathbb{E}\left[\left\|\iint_{\mathcal{X}} f(s,x)q(ds,dx)\right\|^2\right]$$

$$\leq 2\mathbb{E}\left[\left(\iint_{\mathcal{X}} \|f(s,x)\|N(ds,dx)\right)^2\right] + 2\left(\iint_{\mathcal{X}} \|f(s,x)\|\beta(dx)ds\right)^2$$

$$\leq 2\mathbb{E}\left[\left(\iint_{\mathcal{X}} \|f(s,x)\|q(ds,dx) + \iint_{\mathcal{X}} \|f(s,x)\|\beta(dx)ds\right)^2\right]$$

$$+ 2\left(\iint_{\mathcal{X}} \|f(s,x)\|\beta(dx)ds\right)^2$$

$$\leq 4\mathbb{E}\left[\left(\iint_{\mathcal{X}} \|f(s,x)\|q(ds,dx)\right)^2\right] + 6\left(\iint_{\mathcal{X}} \|f(s,x)\|\beta(dx)ds\right)^2.$$

Applying the Itô isometry (3.1.3) yields the desired estimate.   □

The following result complements Theorem 3.1.5.

Theorem 3.1.13 *Let F be a separable Banach space. The following statements are equivalent:*

(a) *The Banach space F is of type 2.*
(b) *There exists a constant $K > 0$ such that for an arbitrary compensated Poisson random measure $q(dt,dx)$ with compensator $\nu(dt,dx) = dt \otimes \beta(dx)$ for some σ-finite measure β, every function $f \in \mathcal{L}_\beta^2(F)$ is Pettis integrable and we have*

$$\mathbb{E}\Big[\Big\|(\text{P}-)\iint_{\mathcal{X}} f(t,x)q(dt,dx)\Big\|^2\Big] \tag{3.1.18}$$

$$\leq K \iint_{\mathcal{X}} \|f(t,x)\|^2 dt\beta(dx) \quad \text{for all } f \in \mathcal{L}^2_\beta(F).$$

Proof The implication (a) $\Rightarrow$ (b) follows from estimate (3.1.15) and Theorem 3.1.5.

In order to prove (b) $\Rightarrow$ (a), we shall establish that $\ell^2_F \subset B_F$, where ℓ^2_F denotes the linear space consisting of all sequences $(x_j)_{j\in\mathbb{N}} \subset F$ with $\sum_{j=1}^{\infty} \|x_j\|^2 < \infty$, and where B_F denotes the linear space of all sequences $(x_j)_{j\in\mathbb{N}} \subset F$ such that for any i.i.d. sequence $(\epsilon_j)_{j\in\mathbb{N}}$ of independent symmetric Bernoulli random variables (i.e. $\mathbb{P}(\epsilon_j = \pm 1) = \frac{1}{2}$) the sequence $(\sum_{j=1}^{n} \epsilon_j x_j)_{n\in\mathbb{N}}$ is bounded in $L^2(\Omega, \mathcal{F}, \mathbb{P}; F)$. Then the Banach space F is of type 2 in the sense of [43, p. 113], and according to [43, Theorem II.6.6] the Banach space F is of type 2 in the sense of our Definition 3.1.6.

Let $(x_j)_{j\in\mathbb{N}} \in \ell^2_F$ be arbitrary. For each $j \in \mathbb{N}$ let N_j be a Poisson random measure on $\mathcal{X}$ with intensity measure $\nu_j(dt, dx) = dt \otimes \beta_j(dx)$, where the measure β_j is given by

$$\beta_j = \frac{1}{2}(\delta_{x_j} + \delta_{-x_j}),$$

and denote by $q_j(dt, dx) = N_j(dt, dx) - \nu_j(dt, dx)$ the associated Poisson random measure. Note that

$$\int_0^1 \int_E x\beta_j(dx)dt = 0, \quad \text{for all } j \in \mathbb{N}. \tag{3.1.19}$$

Let $(\Pi_j)_{j\in\mathbb{N}}$ be an i.i.d. sequence of bilateral Poisson distributed random variables with parameter $\frac{1}{2}$, that is, each Π_j has the distribution of $X-Y$, where X and Y are independent with $X, Y \sim \text{Pois}(\frac{1}{2})$. We define the sequence $(S_n)_{n\in\mathbb{N}_0} \subset L^2(\Omega, \mathcal{F}, \mathbb{P}; F)$ by $S_0 := 0$ and

$$S_n := \sum_{j=1}^n \Pi_j x_j, \quad n \in \mathbb{N}.$$

Then, for all $n, m \in \mathbb{N}_0$ with $m < n$ and all $y^* \in F^*$ we have

$$\mathbb{E}[e^{i\langle y^*, S_n - S_m\rangle}] = \mathbb{E}\Big[\exp\Big(i\Big\langle y^*, \sum_{j=m+1}^n \Pi_j x_j\Big\rangle\Big)\Big] = \prod_{j=m+1}^n \mathbb{E}[e^{i\langle y^*, x_j\rangle \Pi_j}]$$

$$= \prod_{j=m+1}^n \exp\Big(\frac{1}{2}(e^{i\langle y^*, x_j\rangle} - 1)\Big)\exp\Big(\frac{1}{2}(e^{-i\langle y^*, x_j\rangle} - 1)\Big)$$

$$= \exp\left(\frac{1}{2} \sum_{j=m+1}^{n} \left((e^{i\langle y^*, x_j\rangle} - 1) + (e^{-i\langle y^*, x_j\rangle} - 1)\right)\right)$$

$$= \exp\left(\int_0^1 \int_E (e^{i\langle y^*, x\rangle} - 1)\left(\sum_{j=m+1}^{n} \beta_j\right)(dx)dt\right)$$

$$= \mathbb{E}\left[\exp\left(i\langle y^*, \int_0^1 \int_E x\left(\sum_{j=m+1}^{n} N_j\right)(dt, dx)\rangle\right)\right],$$

where we have used Proposition 2.5.3 in the last step. By virtue of the uniqueness theorem for characteristic functions (see Theorem 2.5.1), for all $n, m \in \mathbb{N}_0$ with $m < n$ we obtain

$$S_n - S_m \overset{d}{=} \int_0^1 \int_E x\left(\sum_{j=m+1}^{n} N_j\right)(dt, dx). \qquad (3.1.20)$$

We shall now prove that $(S_n)_{n\in\mathbb{N}}$ is a Cauchy sequence in $L^2(\Omega, \mathcal{F}, \mathbb{P}; F)$. Let $\epsilon > 0$ be arbitrary. Since $\sum_{j=1}^{\infty} \|x_j\|^2 < \infty$, there exists an $n_0 \in \mathbb{N}$ such that for $n, m \geq n_0$ with $m < n$ we have

$$\sum_{j=m+1}^{n} \|x_j\|^2 < \frac{\epsilon}{K}.$$

By Theorem 3.1.5, for every function $f \in \mathcal{L}_\beta^2(F)$ the Wiener integral exists and we have the identity (3.1.8). Using (3.1.8), (3.1.19), (3.1.20) and identity (3.1.17) from Lemma 3.1.11 and estimate (3.1.18), for all $n, m \geq n_0$ with $m < n$ we obtain

$$\mathbb{E}[\|S_n - S_m\|^2] = \mathbb{E}\left[\left\|\int_0^1 \int_E x\left(\sum_{j=m+1}^{n} N_j\right)(dt, dx)\right\|^2\right]$$

$$= \mathbb{E}\left[\left\|\int_0^1 \int_E x\left(\sum_{j=m+1}^{n} q_j\right)(dt, dx)\right\|^2\right]$$

$$= \mathbb{E}\left[\left\|(P-)\int_0^1 \int_E x\left(\sum_{j=m+1}^{n} q_j\right)(dt, dx)\right\|^2\right]$$

$$\leq K \int_0^1 \int_E \|x\|^2 \left(\sum_{j+m+1}^{n} \beta_j(dx)\right)dt = K \sum_{j=m+1}^{n} \|x_j\|^2 < \epsilon.$$

Consequently, the sequence $(S_n)_{n\in\mathbb{N}}$ converges in $L^2(\Omega, \mathcal{F}, \mathbb{P}; F)$. By the Itô–Nisio Theorem, see [46, Theorem 3.1], the sequence $(S_n)_{n\in\mathbb{N}}$ converges almost surely.

We shall now apply the contraction principle from [50]. Note that the sequence $(\Pi_j)_{j\in\mathbb{N}}$ is uniformly nondegenerate, that is, there exist constants $a, b > 0$ such that

$$\mathbb{P}[|\Pi_j| \geq a] \geq b \quad \text{for all } j \in \mathbb{N}.$$

Now let $(\epsilon_j)_{j\in\mathbb{N}}$ be an arbitrary i.i.d. sequence of independent symmetric Bernoulli random variables. According to [50, Theorem 5.6], the series $\sum_{j=1}^{\infty} \epsilon_j x_j$ converges almost surely. Applying the Itô–Nisio Theorem again, the series $\sum_{j=1}^{\infty} \epsilon_j x_j$ converges in $L^2(\Omega, \mathcal{F}, \mathbb{P}; F)$, and hence $(x_j)_{j\in\mathbb{N}} \in B_F$, which settles the proof. $\qquad\square$

3.2 Lévy Processes

Let F be a separable Banach space.

Definition 3.2.1 An F-valued adapted process $X = (X_t)_{t\geq 0}$ with $\mathbb{P}(X_0 = 0) = 1$ is called a *Lévy process* if the following conditions are satisfied:

1. X has independent increments, i.e., $X_t - X_s$ is independent of $\mathcal{F}_s$ for all $0 \leq s \leq t$.
2. X has stationary increments, i.e., $X_t - X_s \stackrel{d}{=} X_{t-s}$ for all $0 \leq s \leq t$.
3. X is continuous in probability, i.e., for all $t \geq 0$ we have $X_t = \lim_{s\to t} X_s$ in probability.
4. With probability 1 the paths $X.(\omega) : \mathbb{R}_+ \to F$ are cádlág.

If 1–3 holds then $X = (X_t)_{t\geq 0}$ is a *Lévy process in law*.

Remark 3.2.2 Let G be another separable Banach space, let $\ell : F \to G$ be a continuous linear operator and $b \in G$. Then the process $Y_t = \ell(X_t) + bt$ is a G-valued Lévy process.

Let us recall that φ_X denotes the characteristic function of the random variable X.

Lemma 3.2.3 *Let X be a Lévy process in law. Then, there exists a function $\psi :$ $F^* \to \mathbb{C}$ such that*

$$\varphi_{X_t}(x^*) = e^{t\psi(x^*)}, \quad (t, x^*) \in \mathbb{R}_+ \times F^*. \tag{3.2.1}$$

Proof Let $x^* \in F^*$ be arbitrary. Then, we have $\varphi_{X_0}(x^*) = 1$ and, since X is continuous in probability, the function $t \mapsto \varphi_{X_t}(x^*)$ is continuous. By the independence and the stationarity of the increments of X, for all $t, s \geq 0$ we obtain

$$\varphi_{X_{t+s}}(x^*) = \mathbb{E}[e^{i\langle x^*, X_{t+s}\rangle}] = \mathbb{E}[e^{\langle x^*, X_{t+s}-X_s+X_s\rangle}]$$

$$= \mathbb{E}[e^{i\langle x^*, X_{t+s}-X_s\rangle}]\mathbb{E}[e^{i\langle x^*, X_s\rangle}]$$

$$= \mathbb{E}[e^{i\langle x^*, X_t\rangle}]\mathbb{E}[e^{i\langle y^*, X_s\rangle}] = \varphi_{X_t}(y^*)\varphi_{X_s}(y^*).$$

It follows that $(\varphi_{X_t}(x^*))_{t\geq 0}$ is a uniform continuous semigroup in $\mathbb{C}$. Therefore, the limit

$$\psi(x^*) := \lim_{t\to 0} \frac{\varphi_{X_t}(x^*) - 1}{t}$$

exists and we have (3.2.1). □

Definition 3.2.4 A function $\psi : F^* \to \mathbb{C}$ is called a *characteristic exponent* of X if we have (3.2.1).

Lemma 3.2.5 *For all $n \in \mathbb{N}$, all $0 = t_0 < \ldots < t_n$ and all $x_1^*, \ldots, x_n^* \in F^*$ we have*

$$\varphi_{(X_{t_1},\ldots,X_{t_n})}(x_1^*, \ldots, x_n^*) = \prod_{k=1}^{n} \varphi_{X_{t_k-t_{k-1}}}\left(\sum_{l=k}^{n} x_l^*\right).$$

Proof Using the stationarity and the independence of the increments, we obtain

$$\varphi_{(X_{t_1},\ldots,X_{t_n})}(x_1^*, \ldots, x_n^*) = \mathbb{E}\left[\exp\left(i\sum_{k=1}^{n}\langle x_k^*, X_{t_k}\rangle\right)\right]$$

$$= \mathbb{E}\left[\exp\left(i\left(\left\langle \sum_{k=1}^{n} x_k^*, X_{t_1}\right\rangle + \sum_{k=2}^{n}\langle x_k^*, X_{t_k} - X_{t_1}\rangle\right)\right)\right]$$

$$= \mathbb{E}\left[\exp\left(i\left\langle \sum_{k=1}^{n} x_k^*, X_{t_1}\right\rangle\right)\right]\mathbb{E}\left[\exp\left(i\sum_{k=2}^{n}\langle x_k^*, X_{t_k} - X_{t_1}\rangle\right)\right]$$

$$= \varphi_{X_{t_1}}(x_1^* + \cdots + x_n^*)\varphi_{(X_{t_2-t_1},\ldots,X_{t_n-t_1})}(x_2^*, \ldots, x_n^*).$$

By induction, the claim follows. □

Lemma 3.2.6 *Let F, G be Banach spaces and let (X, Y) be an $F \times G$-valued Lévy process in law such that for all $t \geq 0$ the random variables X_t and Y_t are independent. Then the two Lévy processes in law X and Y are independent.*

Proof Let $n \in \mathbb{N}$ and $0 \leq t_1 < \ldots < t_n$ be arbitrary. Using Lemma 3.2.5, for all $x_1^*, \ldots, x_n^* \in F^*$ and $y_1^*, \ldots, y_n^* \in G^*$ we obtain

$$\varphi_{(X_{t_1},\ldots,X_{t_n},Y_{t_1},\ldots,Y_{t_n})}(x_1^*, \ldots, x_n^*, y_1^*, \ldots, y_n^*)$$

$$= \varphi_{(X_{t_1},Y_{t_1},\ldots,X_{t_n},Y_{t_n})}(x_1^*, y_1^*, \ldots, x_n^*, y_n^*)$$

$$
= \prod_{k=1}^{n} \varphi_{(X_{t_k-t_{k-1}}, Y_{t_k-t_{k-1}})} \left(\sum_{l=k}^{n} (x_l^*, y_l^*) \right)
$$

$$
= \prod_{k=1}^{n} \varphi_{(X_{t_k-t_{k-1}})} \left(\sum_{l=k}^{n} x_l^* \right) \varphi_{(Y_{t_k-t_{k-1}})} \left(\sum_{l=k}^{n} y_l^* \right)
$$

$$
= \varphi_{(X_{t_1},\ldots,X_{t_n})}(x_1^*, \ldots, x_n^*) \varphi_{(Y_{t_1},\ldots,Y_{t_n})}(y_1^*, \ldots, y_n^*),
$$

proving the independence of the random vectors $(X_{t_1}, \ldots, X_{t_n})$ and $(Y_{t_1}, \ldots, Y_{t_n})$. $\square$

Definition 3.2.7 A measure β on $(F, \mathcal{B}(F))$ is called a *Lévy measure* if it satisfies $\beta(\{0\}) = 0$ and

$$
\int_F (\|x\|^2 \wedge 1) \beta(dx) < \infty. \tag{3.2.2}
$$

Definition 3.2.8 Let X be a Lévy process in law and let β be Lévy measure. We say that β is a *Lévy measure of* X if the function $\psi : F^* \to \mathbb{C}$ given by

$$
\psi(x^*) = \int_F \left(e^{i\langle x^*, x \rangle} - 1 - i\langle x^*, x \rangle \mathbb{1}_{\{\|x\| \leq 1\}} \right) \beta(dx), \quad x^* \in F^*
$$

is a characteristic exponent of X.

Our goal is to construct a Lévy process X with a given Lévy measure β. For this, we prepare some auxiliary results. Let $(E, \mathcal{E})$ be a Blackwell space and let $N(dt, dx)$ be a Poisson random measure on $\mathcal{X} = \mathbb{R}_+ \times E$ with compensator $\nu(dt, dx) = dt \otimes \beta(dx)$. We denote by $q(dt, dx)$ the associated compensated Poisson random measure.

Lemma 3.2.9 *Let* $f \in L^1(E, \mathcal{E}, \beta; F)$. *Then the process*

$$
X_t = \int_0^t \int_E f(x) N(ds, dx), \quad t \geq 0 \tag{3.2.3}
$$

is a Lévy process.

Proof The process X is adapted and we have $\mathbb{P}(X_0 = 0) = 1$. For $t \geq 0$ with $s < t$ we have $f \mathbb{1}_{(0,t] \triangle (0,s]} \to 0$ in $\mathcal{L}^1_\beta(F)$ for $s \to t$ by Lebesgue's dominated convergence theorem. By the continuity of the integral operator (2.4.4) we obtain

$$
X_t - X_s = \iint_{\mathcal{X}} f(x) \mathbb{1}_{(s,t]}(x) N(ds, dx) \to 0
$$

in $L^1(F)$ as $s \to t$, and hence X is continuous in probability.

In order to prove independence and stationarity of the increments of X, we first assume that $f \in \Sigma(F)$ is a simple function of the form

$$f(x) = \sum_{k=1}^{n} a_k \mathbb{1}_{A_k}(x).$$

Then the integral process (3.2.3) is given by

$$X_t = \sum_{k=1}^{n} a_k N((0, t] \times A_k), \quad t \geq 0$$

and is cádlág.
For arbitrary $0 \leq s \leq t$ we obtain

$$X_t - X_s = \sum_{k=1}^{n} a_k N((s, t] \times A_k) \quad \text{and} \quad X_{t-s} = \sum_{k=1}^{n} a_k N((0, t - s] \times A_k).$$

By Definition 2.4.5 and Theorem 2.4.6 we have

$$\mathcal{L}(X_t - X_s) = \text{Pois}_{a_1}((t - s)\beta(A_1)) * \ldots * \text{Pois}_{a_n}((t - s)\beta(A_n)) = \mathcal{L}(X_{t-s}),$$

where $\text{Pois}_a(\lambda)$ denotes a Poisson distribution on the linear space $\langle a \rangle$ with mean λ, and $X_t - X_s$ is independent of $\mathcal{F}_s$. For a general function $f \in L^1(E; F)$ there exists a sequence $(f_n)_{n \in \mathbb{N}}$ of simple functions such that $f_n \to f$ in $L^1(E; F)$. Set

$$X_t^n = \int_0^t \int_E f_n(x) N(ds, dx), \quad t \geq 0.$$

By the continuity of the integral operator (2.4.4) we have $X_t^n \to X_t$ in $L^1(E; F)$ for each $t \geq 0$, and hence X has independent and stationary increments and is cádlág. $\square$

Similarly, the following result can be proven:

Lemma 3.2.10 *Suppose the separable Banach space F is of type 2. Let $f \in L^2(E, \mathcal{E}, \beta; F)$ be arbitrary. Then the process*

$$X_t = \int_0^t \int_E f(x) q(ds, dx), \quad t \geq 0$$

is a Lévy process.

Now, let β be a Lévy measure on $(F, \mathcal{B}(F))$ and let $N(dt, dx)$ be a Poisson random measure on $\mathcal{X} = \mathbb{R}_+ \times F$ with compensator $\nu(dt, dx) = dt \otimes \beta(dx)$. We denote by $q(dt, dx)$ the associated compensated Poisson random measure.

Theorem 3.2.11 *Suppose that the Banach space F is of type 2 or that*

$$\int_F (\|x\| \wedge 1)\beta(dx) < \infty.$$

Then the two processes

$$Y_t := \int_0^t \int_{\{\|x\| \geq 1\}} xN(ds, dx), \quad t \geq 0$$

$$Z_t := \int_0^t \int_{\{\|x\| < 1\}} xq(ds, dx), \quad t \geq 0$$

are independent Lévy processes, and the process $X = Y + Z$ is a Lévy process with Lévy measure β.

Proof For $n \in \mathbb{N}$ we define the process

$$Y_t^n := \int_0^t \int_{\{1 \leq \|x\| \leq n\}} xN(ds, dx), \quad t \geq 0.$$

We can write the process (Y^n, Z) as

$$(Y_t^n, Z_t) = \left(\int_0^t \int_E x\mathbb{1}_{\{1 \leq \|x\| \leq n\}} q(ds, dx) \right.$$

$$\left. + \int_0^t \int_{\{1 \leq \|x\| \leq n\}} x\beta(dx)dt, \int_0^t \int_E x\mathbb{1}_{\{\|x\| < 1\}} q(ds, dx) \right).$$

By Lemmas 2.5.4, 3.1.8, 3.2.6, 3.2.9 and 3.2.10, the process (Y^n, Z) is a Lévy process with independent components. Using Lebesgue's dominated convergence theorem, we have $Y_t^n \to Y_t$ almost surely for every $t \geq 0$. Hence (Y, Z) is a Lévy process with independent components. By Remark 3.2.2, the process $X = Y + Z$ is a Lévy process. Using Propositions 2.5.3 and 3.1.9, for each $x^* \in F^*$ the characteristic function is given by

$$\varphi_X(x^*) = \lim_{n\to\infty} \varphi_{Y_1^n + Z_1}(x^*) = \varphi_{Z_1}(x^*) \lim_{n\to\infty} \varphi_{Y_1^n}(x^*)$$

$$= \exp\left(\int_0^t \int_{\{\|x\|<1\}} \left(e^{i\langle x^*,x\rangle} - 1 - i\langle x^*,x\rangle \right) \beta(dx)ds \right)$$

$$\times \lim_{n\to\infty} \exp\left(\int_0^t \int_{\{1\le\|x\|\le n\}} \left(e^{i\langle x^*,x\rangle} - 1 \right) \beta(dx)ds \right)$$

$$= \exp\left(\int_0^t \int_E \left(e^{i\langle x^*,x\rangle} - 1 - i\langle x^*,x\rangle \mathbb{1}_{\{\|x\|\le 1\}} \right) \beta(dx)ds \right).$$

Consequently, the Lévy process X has the Lévy measure β. $\qquad\qquad\square$

3.3 The Lévy–Itô Decomposition in Banach Spaces

In this section, we prove the Lévy–Itô decomposition in separable Banach spaces, showing that every Lévy process with values in a Banach space can be decomposed into three independent components, which are a drift, a Brownian motion and a jump part, represented by a Wiener integral.

In the sequel, let F be a separable Banach space and let X be an F-valued Lévy process. Set $\mathcal{X} = \mathbb{R}_+ \times F$. We define the random measure N on $(\mathcal{X}, \mathcal{B}(\mathcal{X}))$ by

$$N(A) := \sum_{t>0} \mathbb{1}_{\{\Delta X_t \ne 0\}} \delta_{(t,\Delta X_t)}(A), \quad A \in \mathcal{B}(\mathcal{X}). \tag{3.3.1}$$

Furthermore, for any $B \in \mathcal{B}(F)$ with $0 \notin \bar{B}$ we define the process

$$N_t^B = N((0,t] \times B), \quad t \ge 0. \tag{3.3.2}$$

For $t \ge 0$ let $\mathcal{C}_t$ be the σ-algebra generated by the system of all cylinder sets

$$\{X_{t_1} - X_{t_0} \in B_1, \dots, X_{t_n} - X_{t_{n-1}} \in B_n\}$$

with $n \in \mathbb{N}$, time points $t \le t_0 < t_1 < \dots < t_n$ and Borel sets $B_1, \dots, B_n \in \mathcal{B}(F)$. Note that the σ-algebra $\mathcal{C}_t$ is independent of $\mathcal{F}_t$.

We say that the jumps of a Lévy process X are bounded by a constant $C > 0$ if

$$\mathbb{P}\left(\sup_{t\in\mathbb{R}_+} \|\Delta X_t\| \le C \right) = 1.$$

Lemma 3.3.1 *Let $(X_t)_{t\geq 0}$ be an F-valued Lévy process with bounded jumps. Then we have*

$$\mathbb{E}[\|X_t\|^n] < \infty \quad \text{for all } n \in \mathbb{N} \text{ and } t \geq 0.$$

Proof This is established by literally following the proof of [87, Theorem I.34]. □

Lemma 3.3.2 *For all $0 \leq s < t$ we have that ΔX_t is C_s-measurable.*

Proof We have

$$\Delta X_t = X_t - X_{t-} = \lim_{n\to\infty} (X_t - X_{t-\frac{1}{n}}),$$

which is C_s-measurable. □

Lemma 3.3.3 *For each $B \in \mathcal{B}(F)$ with $0 \notin \overline{B}$ the process N^B is a Lévy process, and the measure β on $(F, \mathcal{B}(F))$ given by*

$$\beta(B) = \mathbb{E}[N_1^B], \quad B \in \mathcal{B}(F) \tag{3.3.3}$$

is σ-finite.

Proof We can write the process N^B as

$$N_t^B = \sum_{0<s\leq t} \mathbb{1}_B(\Delta X_s), \quad t \geq 0.$$

Let $0 \leq s < t$ be arbitrary. By Lemma 3.3.2 we have

$$N_t^B - N_s^B = \sum_{s<u\leq t} \mathbb{1}_B(\Delta X_u)$$

is C_s-measurable, and hence independent of $\mathcal{F}_s$. Define $(\tilde{X}_u)_{u\geq 0}$ by

$$\tilde{X}_u := X_{s+u} - X_s, \quad u \geq 0.$$

Then we have

$$N_t^B - N_s^B = \sum_{s<u\leq t} \mathbb{1}_B(\Delta X_u) = \sum_{0<u\leq t-s} \mathbb{1}_B(\Delta \tilde{X}_u) \quad \text{and}$$

$$N_{t-s}^B = \sum_{0<u\leq t-s} \mathbb{1}_B(\Delta X_u).$$

Consequently, the process N^B has independent and stationary increments, and thus it is a Lévy process. Note that the jumps of N^B are bounded by 1. Lemma 3.3.1 yields that $\beta(B) < \infty$ for all $B \in \mathcal{B}(F)$ with $0 \notin \overline{B}$. Therefore, the measure β is σ-finite. □

Proposition 3.3.4 *The random measure $N(dt, dx)$ is a Poisson random measure with compensator $\nu(dt, dx) = dt \otimes \beta(dx)$, where the measure β is given by (3.3.3).*

Proof Arguing as in the proof of [48, Proposition II.1.16], the random measure N is an integer-valued random measure on $\mathcal{X}$.

Let $s \in \mathbb{R}_+$ and $A \in \mathcal{B}(\mathcal{X})$ with $A \subset (s, \infty) \times E$ and $\mathbb{E}[N(A)] < \infty$ be arbitrary. Then $N(A)$ is $\mathcal{C}_t$-measurable, and hence independent of $\mathcal{F}_s$.

Let $0 \leq s < t$ and $B \in \mathcal{B}(F)$ with $0 \notin \overline{B}$ be arbitrary. Since N^B is a Lévy process by Lemma 3.3.3, we obtain

$$\mathbb{E}[N((s, t] \times B)] = \mathbb{E}[N_t^B - N_s^B] = (t - s)\mathbb{E}[N_1^B] = (t - s)\beta(B).$$

Therefore, N is a Poisson random measure with compensator given by (3.3.3). $\square$

We call a real-valued Lévy process X a *Poisson process* with parameter $\lambda > 0$ if X_1 has a Poisson distribution with mean λ.

Corollary 3.3.5 *For each $B \in \mathcal{B}(E)$ with $0 \notin \overline{B}$ the process N^B is a Poisson process with mean $\beta(B)$.*

Proof This is an immediate consequence of Proposition 3.3.4 and Theorem 2.4.6. $\square$

In the sequel, for a given Lévy process X, the random measure N denotes the Poisson random measure defined in (3.3.1), whose compensator is given by $\nu(dt, dx) = dt \otimes \beta(dx)$ with β defined in (3.3.3), and $q(dt, dx) = N(dt, dx) - \nu(dt, dx)$ denotes the associated compensated Poisson random measure.

Lemma 3.3.6 *Let X be a Lévy process and $B \in \mathcal{B}(E)$ with $0 \notin \overline{B}$. Then the process*

$$\mathcal{W}_t = X_t - \int_0^t \int_B xN(ds, dx)$$

is a Lévy process with

$$\mathbb{P}(\Delta \mathcal{W}_t \notin B) = 1 \quad \text{for all } t \geq 0. \tag{3.3.4}$$

Proof Let $0 \leq s < t$ be arbitrary. Then we have

$$\mathcal{W}_t - \mathcal{W}_s = X_t - X_s - \sum_{s < u \leq t} \Delta X_u \mathbb{1}_B(\Delta X_u) \tag{3.3.5}$$

is $\mathcal{C}_s$-measurable, and hence independent of $\mathcal{F}_s$. Define $(\tilde{X}_u)_{u \geq 0}$ by

$$\tilde{X}_u := X_{s+u} - X_s, \quad u \geq 0.$$

Then we have

$$W_t - W_s = X_t - X_s - \sum_{s<u\leq t} \Delta X_u 1_B(\Delta X_u) = \tilde{X}_{t-s} - \sum_{0<u\leq t-s} \Delta \tilde{X}_u 1_B(\Delta \tilde{X}_u),$$

$$W_{t-s} = X_{t-s} - \sum_{0<u\leq t-s} \Delta X_u 1_B(X_u),$$

hence independent and stationary increments. The representation (3.3.5) also shows that (3.3.4) is valid.   □

Definition 3.3.7 An F-valued process $(W_t)_{t\geq 0}$ is called a Wiener process if for each $n \in \mathbb{N}$ and each $\ell \in \mathcal{L}(E, \mathbb{R}^n)$ the process $\ell(W)$ is an $\mathbb{R}^n$-valued Lévy process with $\ell(W_t) \sim N(0, (t-s)Q_\ell)$ with a symmetric, non-negative definite matrix $Q_\ell \in \mathbb{R}^{n\times n}$.

Proposition 3.3.8 *Let $(X_t)_{t\geq 0}$ be a Lévy process with jumps bounded by 1. Suppose that the Banach space E is of type 2 or that*

$$\int_F (\|x\| \wedge 1)\beta(dx) < \infty. \tag{3.3.6}$$

Then we have $\mathbb{E}[\|X_1\|] < \infty$ and the process $(W_t)_{t\geq 0}$ given by

$$W_t = X_t - t\mathbb{E}[X_1] - \int_0^t \int_{\{\|x\|\leq 1\}} xq(ds, dx), \quad t \geq 0$$

is an F-valued Wiener process.

Proof According to Lemma 3.3.1 we have $\mathbb{E}[\|X_t\|] < \infty$ for all $t \geq 0$. For each $n \in \mathbb{N}$ we define the process

$$W_t^n := X_t - t\mathbb{E}[X_1] - \int_0^t \int_{\{1/n<\|x\|\leq 1\}} xq(ds, dx), \quad t \geq 0.$$

We can write this process as

$$W_t^n = X_t + \left(\int_{\{1/n<\|x\|\leq 1\}} x\beta(dx) - \mathbb{E}[X_1] \right)t - \int_0^t \int_{\{1/n<\|x\|\leq 1\}} xN(ds, dx).$$

If the Banach space F is of type 2, this follows from Lemma 3.1.11. Hence, by Lemma 3.3.6, each W^n is a Lévy process with jumps bounded by $\frac{1}{n}$.

   We fix an arbitrary $T > 0$. If condition (3.3.6) is satisfied, then Doob's inequality (Theorem 2.3.5), Lemma 3.1.12 and Lebesgue's theorem yield

$$\mathbb{E}\left[\sup_{s\in[0,T]}\|W_s - W_s^n\|^2\right]$$

$$= \mathbb{E}\left[\sup_{s\in[0,T]}\left\|\int_0^s\int_{\{\|x\|<\frac{1}{n}\}} xq(du, dx)\right\|^2\right]$$

$$\leq 4\sup_{s\in[0,T]}\mathbb{E}\left[\left\|\int_0^s\int_{\{\|x\|<\frac{1}{n}\}} xq(du, dx)\right\|^2\right]$$

$$\leq 16\int_0^T\int_{\{\|x\|<\frac{1}{n}\}}\|x\|^2\beta(dx)ds + 24\left(\int_0^T\int_{\{\|x\|<\frac{1}{n}\}}\|x\|\beta(dx)ds\right)^2$$

$$\leq 16T\int_{\{\|x\|<\frac{1}{n}\}}\|x\|\beta(dx) + 24T^2\left(\int_{\{\|x\|<\frac{1}{n}\}}\|x\|\beta(dx)\right)^2 \to 0.$$

In the other case, where the separable Banach space E is of type 2, Doob's inequality (Theorem 2.3.5), estimate (3.1.15) and Lebesgue's theorem give us

$$\mathbb{E}\left[\sup_{s\in[0,T]}\|W_s - W_s^n\|^2\right] = \mathbb{E}\left[\sup_{s\in[0,T]}\left\|\int_0^s\int_{\{\|x\|<\frac{1}{n}\}} xq(du, dx)\right\|^2\right]$$

$$\leq 4\sup_{s\in[0,T]}\mathbb{E}\left[\left\|\int_0^s\int_{\{\|x\|<\frac{1}{n}\}} xq(du, dx)\right\|^2\right]$$

$$\leq 4K\int_0^T\int_{\{\|x\|<\frac{1}{n}\}}\|x\|^2\beta(dx)ds$$

$$= 4TK\int_{\{\|x\|<\frac{1}{n}\}}\|x\|^2\beta(dx) \to 0.$$

Consequently, there exists a subsequence $(n_k)_{k\in\mathbb{N}}$ such that

$$\sup_{s\in[0,T]}\|W_s - W_s^{n_k}\|^2 \to 0 \quad \mathbb{P}\text{-almost surely.}$$

Since $T > 0$ was arbitrary, it follows that the sample paths of W are continuous. It follows that W is a Wiener process. $\qquad\square$

Lemma 3.3.9 *Let J be a pure jump Lévy process with some Lévy measure β and let W be a Wiener process. Then the two processes W and J are independent.*

Proof By Proposition 2.1.1 it suffices to show that for all $n \in \mathbb{N}$ and $x^* \in L(F, \mathbb{R}^n)$ the processes $\langle x^*, W \rangle$ and $\langle x^*, J \rangle$ are independent.

Let $n \in \mathbb{N}$ and $x^* \in L(F, \mathbb{R}^n)$ be arbitrary. The process $(\langle x^*, W \rangle, \langle x^*, J \rangle)$ is an $\mathbb{R}^{2n}$-valued Lévy process. Indeed, the semimartingale characteristics (see [48, Definition II.2.7]) of $\langle x^*, W \rangle$ and $\langle x^*, J \rangle$ with respect to the truncation function $h : \mathbb{R}^n \to \mathbb{R}^n$, $h(x) = x \mathbb{1}_{\{\|x\| \leq 1\}}$ are given by

$$B^{\langle x^*, W \rangle} \equiv 0, \ C_t^{\langle x^*, W \rangle} = ct, \ \nu^{\langle x^*, W \rangle} \equiv 0,$$
$$B^{\langle x^*, J \rangle} \equiv 0, \ C^{\langle x^*, J \rangle} \equiv 0, \ \nu^{\langle x^*, J \rangle}(A \times B) = \lambda(A)F(B)$$

for some symmetric, non-negative-definite matrix $c \in \mathbb{R}^{n \times n}$ and a measure F on $(\mathbb{R}^n, \mathcal{B}(\mathbb{R}^n))$ satisfying $\int_{\mathbb{R}^n} (\|x\|^2 \wedge 1) F(dx) < \infty$. Hence, we compute the semimartingale characteristics of $(\langle x^*, W \rangle, \langle x^*, J \rangle)$ as

$$B \equiv 0, \quad C_t = \begin{pmatrix} c & 0 \\ 0 & 0 \end{pmatrix} t, \quad \nu(A \times B \times C) = \lambda(A)F(C),$$

showing that $(\langle x^*, W \rangle, \langle x^*, J \rangle)$ is a Lévy process. Computing the characteristic functions yields

$$\varphi_{(\langle x^*, W_1 \rangle, \langle x^*, J_1 \rangle)}(u, v) = \exp\left(\frac{c}{2}|u|^2 + \int_{\mathbb{R}^n} \left(e^{i\langle v, x \rangle} - 1 - i\langle v, x \rangle \, {}_{\{\|x\| \leq 1\}}\right) F(dx)\right)$$

$$= \exp\left(\frac{c}{2}|u|^2\right) \exp\left(\int_{\mathbb{R}^n} \left(e^{i\langle v, x \rangle} - 1 - i\langle v, x \rangle \, {}_{\{\|x\| \leq 1\}}\right) F(dx)\right) = \varphi_{\langle x^*, W_1 \rangle}(u) \varphi_{\langle x^*, J_1 \rangle}(v)$$

for all $(u, v) \in \mathbb{R}^{2n}$. This proves the independence of $\langle x^*, W \rangle$ and $\langle x^*, J \rangle$. $\quad\square$

Theorem 3.3.10 *Let $(X_t)_{t \geq 0}$ be a Lévy process. Suppose that the Banach space F is of type 2 or that*

$$\int_F (\|x\| \wedge 1)\beta(dx) < \infty.$$

Then, the process $(J_t)_{t \geq 0}$ defined as

$$J_t := \int_0^t \int_{\{\|x\| \geq 1\}} xN(ds, dx) + \int_0^t \int_{\{\|x\| < 1\}} xq(ds, dx), \quad t \geq 0$$

is a Lévy process with Lévy measure β, where β is defined in Definition 3.2.8, and there exist $\alpha \in F$ and a Wiener process W such that we have the decomposition

$$X_t = \alpha t + W_t + J_t, \quad t \geq 0. \tag{3.3.7}$$

Moreover, the two processes W and J are independent.

Proof Applying Theorem 3.2.11 yields that J is a Lévy process with Lévy measure β. We define the process L as

$$L_t = \int_0^t \int_{\{\|x\| \geq 1\}} x N(dt, dx), \quad t \geq 0.$$

By Lemmas 3.2.9 and 3.3.6 the processes L and $X - L$ are Lévy processes, and the jumps of $X - L$ are bounded by 1. Hence, by Proposition 3.3.8 we have $\mathbb{E}[\|X_1 - L_1\|] < \infty$, and the process

$$W_t := X_t - L_t - \alpha t - \int_0^t \int_{\{\|x\| < 1\}} x q(ds, dx), \quad t \geq 0$$

is a Wiener process, where $\alpha = \mathbb{E}[X_1 - L_1]$, which provides the Lévy–Itô decomposition (3.3.7). According to Lemma 3.3.9, the two processes W and J are independent. $\qquad\square$

3.4 Isomorphisms for Spaces of Predictable Processes

In order to establish our subsequent results concerning stochastic integration of adapted, measurable resp. progressively measurable processes, we provide the following Theorem 3.4.2 in this section.

Let $(\tilde{\Omega}, \tilde{\mathbb{P}}, \tilde{\mathcal{F}})$ be a measure space. In view of our applications in Sect. 3.5, we do not demand that $(\tilde{\Omega}, \tilde{\mathbb{P}}, \tilde{\mathcal{F}})$ is a probability space. Moreover, let $(\tilde{\mathcal{F}}_t)_{t \geq 0}$ be a filtration satisfying the usual conditions.

Fix $T > 0$ and let μ be a measure on $(\tilde{\Omega} \times [0, T], \tilde{\mathcal{F}}_T \otimes \mathcal{B}([0, T]))$ with marginals

$$\mu(A \times [0, T]) = \tilde{\mathbb{P}}(A), \quad A \in \tilde{\mathcal{F}}_T.$$

We assume that there exists a sequence $(A_n)_{n \in \mathbb{N}} \subset \tilde{\mathcal{F}}_0$ such that $A_n \uparrow \tilde{\Omega}$ and $\tilde{\mathbb{P}}(A_n) < \infty$ for all $n \in \mathbb{N}$. In particular, the measures $\tilde{\mathbb{P}}$ and μ are σ-finite.

There exists a transition kernel $K : \tilde{\Omega} \times \mathcal{B}([0, T]) \to \mathbb{R}_+$ from $(\tilde{\Omega}, \tilde{\mathcal{F}}_T)$ to $([0, T], \mathcal{B}([0, T]))$ such that

$$\mu(B) = \int\limits_{\tilde{\Omega}} \int\limits_{0}^{T} \mathbb{1}_B(\tilde{\omega}, t) K(\tilde{\omega}, dt) \tilde{\mathbb{P}}(d\tilde{\omega}), \quad B \in \tilde{\mathcal{F}}_T \otimes \mathcal{B}([0, T]),$$

see [48, Sect. II.1a].

Definition 3.4.1 We denote by $\tilde{\mathcal{P}}_T$ the predictable σ-algebra on $\tilde{\Omega} \times [0, T]$. Fixing an arbitrary $p \geq 1$, we define the spaces

$$L^p_{T,\text{pred}}(F) := L^p(\tilde{\Omega} \times [0, T], \tilde{\mathcal{P}}_T, \mu; F),$$

$$L^p_{T,\text{prog}}(F) := L^p(\tilde{\Omega} \times [0, T], \tilde{\mathcal{F}}_T \otimes \mathcal{B}([0, T]), \mu; F) \cap \text{Prog}_T(F),$$

$$L^p_{T,\text{ad}}(F) := L^p(\tilde{\Omega} \times [0, T], \tilde{\mathcal{F}}_T \otimes \mathcal{B}([0, T]), \mu; F) \cap \text{Ad}_T(F),$$

where $\text{Prog}_T(F)$ denotes the linear space of all F-valued progressively measurable processes $(\Phi_t)_{t \in [0,T]}$ and $\text{Ad}_T(F)$ denotes the linear space of all F-valued adapted processes $(\Phi_t)_{t \in [0,T]}$.

We have the inclusions

$$L^p_{T,\text{pred}}(F) \subset L^p_{T,\text{prog}}(F) \subset L^p_{T,\text{ad}}(F).$$

In the upcoming theorem, we will show that these three spaces are actually isometrically isomorphic, provided the measures $A \mapsto K(\omega, A)$ are absolutely continuous. In particular, the latter two spaces are Banach spaces, too.

Theorem 3.4.2 *Suppose there is a nonnegative, measurable function $f : \tilde{\Omega} \times [0, T] \to \mathbb{R}$ such that for each $\tilde{\omega} \in \tilde{\Omega}$ we have $K(\tilde{\omega}, dt) = f(\tilde{\omega}, t)dt$. Then we have*

$$L^p_{T,\text{pred}}(F) \cong L^p_{T,\text{prog}}(F) \cong L^p_{T,\text{ad}}(F).$$

Proof It suffices to prove that for each $\Phi \in L^p_{T,\text{ad}}(F)$ there exists a process $\pi(\Phi) \in L^p_{T,\text{pred}}(F)$ such that $\Phi = \pi(\Phi)$ almost everywhere with respect to μ.

Let $\Phi \in L^p_{T,\text{ad}}(F)$ be arbitrary. We will show that there is a sequence $(\Phi^n)_{n \in \mathbb{N}} \subset L^p_{T,\text{pred}}(F)$ such that $\Phi^n \to \Phi$ in $L^p_{T,\text{ad}}(F)$. Then $(\Phi^n)_{n \in \mathbb{N}}$ is a Cauchy sequence in $L^p_{T,\text{pred}}(F)$ and thus has a limit $\pi(\Phi) \in L^p_{T,\text{pred}}(F)$. But this limit has the property $\Phi = \pi(\Phi)$ almost everywhere with respect to μ, which will finish the proof.

The proof of the existence of a sequence $(\Phi^n)_{n \in \mathbb{N}} \subset L^p_{T,\text{pred}}(F)$ satisfying $\Phi^n \to \Phi$ in $L^p_{T,\text{ad}}(F)$ is divided into two steps:

1. First of all, we may assume that

$$\tilde{\mathbb{P}}(\tilde{\Omega}) = \mu(\tilde{\Omega} \times [0, T]) < \infty \tag{3.4.1}$$

and that there is a constant $M > 0$ such that

$$\|\Phi\| \le M \quad \text{everywhere.} \tag{3.4.2}$$

Indeed, by assumption, there exists a sequence $(A_n)_{n\in\mathbb{N}} \subset \tilde{\mathcal{F}}_0$ with $A_n \uparrow \tilde{\Omega}$ and $\tilde{\mathbb{P}}(A_n) < \infty$ for all $n \in \mathbb{N}$. Defining the sequence $(\Phi^n)_{n\in\mathbb{N}} \subset L^p_{T,\text{ad}}(F)$ by $\Phi^n := (\Phi \wedge n)\mathbb{1}_{A_n}$, Lebesgue's dominated convergence theorem yields that $\Phi^n \to \Phi$ in $L^p_{T,\text{ad}}(F)$.

2. Now we proceed with a similar technique as in [58, pp. 97–99]. We extend Φ to a process $(\Phi_t)_{t\in\mathbb{R}}$ by setting

$$\Phi_t(\tilde{\omega}) := 0 \quad \text{for}(\tilde{\omega}, t) \in \tilde{\Omega} \times \mathbb{R} \setminus [0, T].$$

Defining for $n \in \mathbb{N}$ the function $\theta_n : \mathbb{R} \to \mathbb{R}$ by

$$\theta_n(t) := \sum_{j\in\mathbb{Z}} \frac{j-1}{2^n}\mathbb{1}_{(\frac{j-1}{2^n}, \frac{j}{2^n}]}(t),$$

we have $\theta_n(t) \uparrow t$ for all $t \in \mathbb{R}$. The shift semigroup $(S_t)_{t\ge0}$, $S_t f = f(t + \cdot)$ is strongly continuous on $L^p(\mathbb{R}; F)$. Thus, performing integration by the substitution $t \rightsquigarrow t + s$, using Fubini's theorem, Lebesgue's dominated convergence theorem and noting (3.4.1) and (3.4.2) we obtain

$$\int_{\tilde{\Omega}} \int_0^T \int_0^T \|\Phi_{s+\theta_n(t-s)}(\tilde{\omega}) - \Phi_t(\tilde{\omega})\|^p \, dt \, ds \, \tilde{\mathbb{P}}(d\tilde{\omega})$$

$$= \int_{\tilde{\Omega}} \int_0^T \int_{-s}^{T-s} \|\Phi_{s+\theta_n(t)}(\tilde{\omega}) - \Phi_{s+t}(\tilde{\omega})\|^p \, dt \, ds \, \tilde{\mathbb{P}}(d\tilde{\omega})$$

$$= \int_{\tilde{\Omega}} \int_0^T \int_0^{T-t} \|\Phi_{s+\theta_n(t)}(\tilde{\omega}) - \Phi_{s+t}(\tilde{\omega})\|^p \, ds \, dt \, \tilde{\mathbb{P}}(d\tilde{\omega})$$

$$+ \int_{\tilde{\Omega}} \int_{-T}^0 \int_{-t}^T \|\Phi_{s+\theta_n(t)}(\tilde{\omega}) - \Phi_{s+t}(\tilde{\omega})\|^p \, ds \, dt \, \tilde{\mathbb{P}}(d\tilde{\omega}) \to 0.$$

After passing to a subsequence, if necessary, for $\tilde{\mathbb{P}} \otimes \lambda \otimes \lambda$-almost all $(\tilde{\omega}, s, t) \in \tilde{\Omega} \times [0, T] \times [0, T]$ we have

$$\|\Phi_{s+\theta_n(t-s)}(\tilde{\omega}) - \Phi_t(\tilde{\omega})\|^p \to 0,$$

where λ denotes the Lebesgue measure. Thus, there exists an $s \in [0, T]$ such that

$$\|\Phi_{s+\theta_n(t-s)}(\tilde{\omega}) - \Phi_t(\tilde{\omega})\|^p \to 0 \quad \text{for} \quad \tilde{\mathbb{P}} \otimes \lambda\text{-almost all } (\tilde{\omega}, t) \in \tilde{\Omega} \times [0, T].$$

$$(3.4.3)$$

For $n \in \mathbb{N}$ we define the process $\Phi^n = (\Phi_t^n)_{t\in[0,T]}$ by

$$\Phi_t^n := \Phi_{s+\theta_n(t-s)} = \sum_{j\in\mathbb{Z}} \Phi_{s+\frac{i-1}{2^n}} \mathbb{1}_{(s+\frac{i-1}{2^n}, s+\frac{j}{2^n}]}(t), \quad t \in [0, T].$$

Note that Φ^n is predictable, because Φ is adapted. Hence, we have $(\Phi^n)_{n\in\mathbb{N}} \subset L_{T,\text{pred}}^p(F)$. By assumption, there is a nonnegative, measurable function $f : \tilde{\Omega} \times [0, T] \to \mathbb{R}$ such that for each $\tilde{\omega} \in \tilde{\Omega}$ we have $K(\tilde{\omega}, dt) = f(\tilde{\omega}, t)dt$. Using (3.4.1) we have

$$\int_{\tilde{\Omega}} \int_0^T f(\tilde{\omega}, t)dt\tilde{\mathbb{P}}(d\tilde{\omega}) = \int_{\tilde{\Omega}} \int_0^T K(\tilde{\omega}, dt)\tilde{\mathbb{P}}(d\tilde{\omega}) = \mu(\tilde{\Omega} \times [0, T]) < \infty.$$

Noting (3.4.1) and (3.4.2), we obtain by (3.4.3) and Lebesgue's dominated convergence theorem

$$\iint_{\tilde{\Omega}\times[0,T]} \|\Phi^n - \Phi\|^p d\mu = \int_{\tilde{\Omega}} \int_0^T \|\Phi_{s+\theta_n(t-s)}(\tilde{\omega}) - \Phi_t(\tilde{\omega})\|^p K(\tilde{\omega}, dt)\tilde{\mathbb{P}}(d\tilde{\omega})$$

$$= \int_{\tilde{\Omega}} \int_0^T \|\Phi_{s+\theta_n(t-s)}(\tilde{\omega}) - \Phi_t(\tilde{\omega})\|^p f(\tilde{\omega}, t)dt\tilde{\mathbb{P}}(d\tilde{\omega}) \to 0,$$

showing that $\Phi^n \to \Phi$ in $L_{T,\text{ad}}^p(F)$. $\square$

3.5 The Itô Integral

Let $(E, \mathcal{E})$ be a Blackwell space and let $q(dt, dx)$ be a compensated Poisson random measure on $\mathbb{R}_+ \times E$ with compensator $\nu(dt, dx) = dt \otimes \beta(dx)$. Let F be a separable Banach space.

Let $I \subset \mathbb{R}_+$ be an index set such that $I = \mathbb{R}_+$ or $I = [0, T]$ for some $T > 0$. A function $f : \Omega \times I \times E \to F$ is called adapted/measurable/progressively measurable if it has the respective properties on the enlarged space

$$(\tilde{\Omega}, \tilde{\mathcal{F}}, \tilde{\mathcal{F}}_t, \tilde{\mathbb{P}}) = (\Omega \times E, \mathcal{F} \times \mathcal{E}, (\mathcal{F}_t \times \mathcal{E})_{t\geq 0}, \mathbb{P} \otimes \beta).$$

Now we fix $T > 0$ and set $\mu = \tilde{\mathbb{P}} \otimes \lambda$. Let $\mathcal{L}^2_{T,\beta}(F) = L^2_{T,\mathrm{ad}}(F)$ and $\Sigma_T(F) = \Sigma(\tilde{\mathcal{G}}_T; F)$. By Proposition 2.1.6 and Theorem 3.4.2 the linear space $\Sigma_T(F)$ is dense in $\mathcal{L}^2_{T,\beta}(F)$. Note that any function $f \in \Sigma_T(F)$ is of the form

$$f(t, x) = \sum_{k=1}^{n} \sum_{l=1}^{m} a_{k,l} \mathbb{1}_{A_{k,l}}(x) \mathbb{1}_{F_{k,l}} \mathbb{1}_{(t_{k-1}, t_k]}(t) \tag{3.5.1}$$

for $n, m \in \mathbb{N}$ with:

- elements $a_{k,l} \in F$ for $k = 1, \ldots, n$ and $l = 1, \ldots, m$;
- time points $0 \le t_0 \le \ldots \le t_n \le T$;
- sets $A_{k,l} \in \mathcal{E}$ with $\beta(A_{k,l}) < \infty$ for $k = 1, \ldots, n$ and $l = 1, \ldots, m$ such that the product sets $A_{k,l} \times (t_{k-1}, t_k]$ are mutually disjoint;
- sets $F_{k,l} \in \mathcal{F}_{t_{k-1}}$ for $k = 1, \ldots, n$ and $l = 1, \ldots, m$.

For $f \in \Sigma_T(F)$ we define the *Itô integral* as the process

$$\int_0^t \int_E f(s, x) q(ds, dx) := \sum_{k=1}^{n} \sum_{l=1}^{m} a_{k,l} \mathbb{1}_{F_{k,l}} q((t_{k-1}, t_k] \cap (0, t] \times A_{k,l}), \quad t \in [0, T].$$

$$\tag{3.5.2}$$

Remark 3.5.1 Assume that the compensated Poisson random measure $q(dt, dx)$ is the counting measure of a Lévy process $(X_t)_{t \ge 0}$, then for $f \in \Sigma_T(F)$

$$\int_0^t \int_E f(s, x) q(ds, dx) = \sum_{0 < s \le t} f(s, (\Delta X_s)(\omega), \omega) \mathbb{1}_E(\Delta X_s(\omega))$$

$$- \int_0^t \int_E f(s, x, \omega) \nu(ds, dx). \tag{3.5.3}$$

Lemma 3.5.2 *For each $f \in \Sigma_T(F)$ the process $M = (M_t)_{t \ge 0}$ given by*

$$M_t = \int_0^t \int_E f(s, x) q(ds, dx), \quad t \in [0, T]$$

belongs to $\mathcal{M}^2_T(F)$.

Proof This is a direct consequence of Lemma 2.4.8. $\qquad\square$

Due to Lemma 3.5.2, we may regard the Itô integral defined in (3.5.2) as a linear operator

$$\Sigma_T(F) \to \mathcal{M}_T^2(F), \quad f \mapsto \left(\int_0^t \int_E f(s, x) q(ds, dx) \right)_{t \in [0,T]}. \tag{3.5.4}$$

Remark 3.5.3 If $F = H$ is a separable Hilbert space, then for simple functions we have the so-called Itô isometry

$$\mathbb{E}\left[\left\| \int_0^T \int_E f(s, x) q(ds, dx) \right\|^2 \right]$$

$$= \mathbb{E}\left[\int_0^T \int_E \|f(s, x)\|^2 \beta(dx) ds \right] \quad \text{for all} \in \mathcal{L}_{T,\beta}^2(H). \tag{3.5.5}$$

Indeed, for a simple $f \in \Sigma_T(H)$ of the form (3.5.1) we have

$$\mathbb{E}\left[\left\| \int_0^T \int_E f(t, x) q(dt, dx) \right\|^2 \right]$$

$$= \mathbb{E}\left[\left\| \sum_{k=1}^n \sum_{l=1}^m a_{k,l} \mathbb{1}_{F_{k,l}} q((t_{k-1}, t_k] \times A_{k,l}) \right\|^2 \right]$$

$$= \mathbb{E}\left[\left\langle \sum_{k=1}^n \sum_{l=1}^m a_{k,l} \mathbb{1}_{F_{k,l}} q((t_{k-1}, t_k] \times A_{k,l}), \sum_{k=1}^n \sum_{l=1}^m a_{k,l} \mathbb{1}_{F_{k,l}} q((t_{k-1}, t_k] \times A_{k,l}) \right\rangle \right]$$

$$= \sum_{k=1}^n \sum_{l=1}^m \|a_{k,l}\|^2 \mathbb{E}[\mathbb{1}_{F_{k,l}} q((t_{k-1}, t_k] \times A_{k,l})]$$

$$+ \sum_{k=1}^n \sum_{l=1}^m \sum_{i=1}^n \sum_{j=1}^m a_{k,l} a_{i,j} \mathbb{E}[\langle \mathbb{1}_{F_{k,l}} q((t_{k-1}, t_k] \times A_{k,l}), \mathbb{1}_{F_{i,j}} q((t_{i-1}, t_i] \times A_{i,j}) \rangle].$$

Using Theorem 2.4.6 we obtain

$$\sum_{k=1}^n \sum_{l=1}^m \|a_{k,l}\|^2 \mathbb{E}[\mathbb{1}_{F_{k,l}} q((t_{k-1}, t_k] \times A_{k,l})]$$

$$= \sum_{k=1}^n \sum_{l=1}^m \|a_{k,l}\|^2 \mathbb{P}(F_{k,l}) \beta(A_{k,l}) \lambda((t_{k-1}, t_k])$$

$$= \mathbb{E}\left[\int_0^T \int_E \|f(t, x)\|^2 \beta(dx) dt \right].$$

·For $k < i$ the random variable $q((t_{i-1}, t_i] \times A_{i,j})$ is independent of $\mathcal{F}_{t_{i-1}}$ and $q((t_{k-1}, t_k] \times A_{k,l})$ is $\mathcal{F}_{t_{i-1}}$-measurable. Therefore, we get

$$\mathbb{E}[\langle \mathbb{1}_{F_{k,l}} q((t_{k-1}, t_k] \times A_{k,l}), \mathbb{1}_{F_{i,j}} q((t_{i-1}, t_i] \times A_{i,j}) \rangle]$$
$$= \mathbb{E}[\mathbb{1}_{F_{k,l}} \mathbb{1}_{F_{i,j}} \mathbb{E}[\langle q((t_{k-1}, t_k] \times A_{k,l}), q((t_{i-1}, t_i] \times A_{i,j}) \rangle] \mid \mathcal{F}_{t_{i-1}}]$$
$$= \mathbb{E}[\mathbb{1}_{F_{k,l}} \mathbb{1}_{F_{i,j}} \langle q((t_{k-1}, t_k] \times A_{k,l}), \mathbb{E}[q((t_{i-1}, t_i] \times A_{i,j})] \rangle] = 0,$$

and hence, the Itô isometry (3.5.6) is valid.

Therefore, if $F = H$ is a separable Hilbert space, then the integral operator (3.5.4) is an isometry, and therefore in particular continuous. Thus, and because $\Sigma_T(H)$ is dense in $\mathcal{L}^2_{T,\beta}(H)$, it has a unique extension

$$\mathcal{L}^2_{T,\beta}(H) \to \mathcal{M}^2_T(H), \quad f \mapsto \left(\int_0^t \int_E f(s, x) q(ds, dx) \right)_{t \in [0,T]},$$

which we call the *Itô integral*, and we have the Itô isometry

$$\mathbb{E}\left[\left\| \int_0^T \int_E f(s, x) q(ds, dx) \right\|^2 \right]$$

$$= \mathbb{E}\left[\int_0^T \int_E \|f(s, x)\|^2 \beta(dx) ds \right] \quad \text{for all } f \in \mathcal{L}^2_{T,\beta}(H). \tag{3.5.6}$$

In order to define the Itô integral in the general setting, where the separable Banach space F is not a Hilbert space, we assume there exists a constant $K_\beta > 0$ (which may depend on β), such that

$$\mathbb{E}\left[\left\| \int_0^T \int_E f(s, x) q(ds, dx) \right\|^2 \right]$$

$$\leq K_\beta \mathbb{E}\left[\int_0^T \int_E \|f(s, x)\|^2 ds \beta(dx) \right] \quad \text{for all } f \in \Sigma(F). \tag{3.5.7}$$

In this case, we can analogously define the Itô integral for all $f \in \mathcal{L}^2_{T,\beta}(F)$ as the continuous linear operator

$$\mathcal{L}^2_{T,\beta}(F) \to \mathcal{M}^2_T(F), \quad f \mapsto \left(\int_0^t \int_E f(s, x) q(ds, dx) \right)_{t \in [0,T]}, \tag{3.5.8}$$

which is the unique extension of (3.5.4). In particular, we obtain the estimate

$$\mathbb{E}\left[\left\|\int\limits_0^T\int\limits_E f(s,x)q(ds,dx)\right\|^2\right]$$

$$\leq K_\beta \mathbb{E}\left[\int\limits_0^T\int\limits_E \|f(s,x)\|^2\beta(dx)ds\right] \quad \text{for all} f \in \mathcal{L}^2_{T,\beta}(F). \tag{3.5.9}$$

Remark 3.5.4 The Itô integral in (3.5.8) is càdlàg.

We proceed with the definition of the Pettis integral. Let $f : \Omega \times [0, T] \times E \to F$ be a function such that $\langle y^*, f\rangle \in \mathcal{L}^2_{T,\beta}(\mathbb{R})$ for all $y^* \in F^*$. We define the linear operator

$$T^f : F^* \to \mathcal{M}^2_T(\mathbb{R}), \quad T^f y^* := \left(\int\limits_0^t\int\limits_E \langle y^*, f(s,x)\rangle q(ds,dx)\right)_{t\in[0,T]}.$$

Arguing as in the proof of Dunford's lemma (see Lemma 2.1.7), we show that T^f is continuous. The function f is called *Pettis integrable* if there exists a process $Z^f \in \mathcal{M}^2_T(F)$ such that almost surely

$$T^f y^* = \langle y^*, Z^f\rangle.$$

Note that such a process Z^f is $\mathbb{P}$-almost surely unique, provided it exists, and that the set of Pettis integrable functions forms a linear space. Following the ideas of [90], we call Z^f the *Pettis integral* of f and set

$$(\text{P}-)\int\limits_0^t\int\limits_E f(s,x)q(ds,dx) := Z^f_t, \quad t \in [0,T].$$

We observe that for each simple function $f \in \Sigma_T(F)$ the Pettis integral exists and coincides with the Itô integral, that is, we have

$$\left(\int\limits_0^t\int\limits_E f(s,x)q(ds,dx)\right)_{t\in[0,T]}$$

$$= \left((\text{P}-)\int\limits_0^t\int\limits_E f(s,x)q(ds,dx)\right)_{t\in[0,T]} \quad \text{for all} f \in \Sigma_T(F).$$

Arguing as in the proof of Lemma 3.1.3, we obtain the following result:

Lemma 3.5.5 *Suppose there exists a constant $K_\beta > 0$ such that (3.5.7) is satisfied. Then, each function $f \in \mathcal{L}_\beta^2(F)$ is Pettis integrable and we have*

$$\iint_{\mathcal{X}} f(s,x)q(ds,dx) = (P-)\iint_{\mathcal{X}} f(s,x)q(ds,dx) \quad \text{for all } f \in \mathcal{L}_\beta^2(F). \quad (3.5.10)$$

Moreover, for each $x^* \in F^*$ we have

$$\left\langle x^*, \iint_{\mathcal{X}} f(s,x)q(ds,dx) \right\rangle = \iint_{\mathcal{X}} \langle x^*, f(s,x)\rangle q(ds,dx) \quad \text{for all } f \in \mathcal{L}_\beta^2(F).$$

$$(3.5.11)$$

Remark 3.5.6 Let us stress that Lemma 3.5.5 states in particular that if there exists a constant $K_\beta > 0$ such that (3.5.7) is satisfied, then the Itô integral is defined on $\mathcal{L}_\beta^2(F)$ and inequality (3.5.9) holds.

Now the question arises which properties of the continuous linear operator T^f ensure that the function $f : \Omega \times [0, T] \times E \to F$ is Pettis integrable. Arguing as in the proof of Theorem 3.1.4, we obtain the following result:

Theorem 3.5.7 *Suppose that $F = H$ is a separable Hilbert space, and let $f : \Omega \times [0, T] \times E \to H$ be a function with $\langle y, f \rangle \in \mathcal{L}_{T,\beta}^2(\mathbb{R})$ for all $y \in H$. Then the following statements are equivalent:*

1. *$T^f : H \to \mathcal{M}_T^2(\mathbb{R})$ is a Hilbert–Schmidt operator.*
2. *$f \in \mathcal{L}_{T,\beta}^2(H)$.*
3. *The function f is Pettis integrable with*

$$\mathbb{E}\left[\left\| (P-) \int_0^T \int_E f(s,x)q(ds,dx) \right\|^2 \right] < \infty. \quad (3.5.12)$$

If F is a separable Banach space, we obtain the following result by arguing as in the proof of Theorem 3.1.5.

Theorem 3.5.8 *Let E be a separable Banach space. The following statements are equivalent:*

1. *Each $f \in \mathcal{L}_{T,\beta}^2(F)$ is Pettis integrable and we have (3.5.12).*
2. *Each $f \in \mathcal{L}_{T,\beta}^2(F)$ is Pettis integrable, we have (3.5.12) and the linear operator*

$$\mathcal{L}_{T,\beta}^2(F) \to \mathcal{M}_T^2(F), \quad f \mapsto (P-) \int_0^T \int_E f(s,x)q(ds,dx) \quad (3.5.13)$$

is continuous.

3. *There exists a constant $K_\beta > 0$ such that (3.5.7) is satisfied.*

If the previous conditions are fulfilled, then we have identities (3.5.10) and (3.5.11).

Now the natural question arises if for non-Hilbertian Banach spaces there exists a constant $K_\beta > 0$ such that inequality (3.5.7) is satisfied for all $f \in \Sigma_T(F)$. Following [85], we introduce M-type 2 Banach spaces for this purpose.

Definition 3.5.9 A separable Banach space F is called a *Banach space of M-type 2* if there exists a constant $K > 0$ such that for each $n \in \mathbb{N}$, for every filtered probability space $(\Omega, \mathcal{F}, (\mathcal{F}_i)_{i=0,\ldots,n}, \mathbb{P})$ and for every F-valued $(\mathcal{F}_k)$-martingale $(M_i)_{i=0,\ldots,n}$ with $M_0 = 0$ we have

$$\mathbb{E}[\|M_n\|^2] \leq K \sum_{i=1}^{n} \mathbb{E}[\|M_i - M_{i-1}\|^2]. \tag{3.5.14}$$

Remark 3.5.10 Note that every Banach space of M-type 2 is also a Banach space of type 2. Indeed, for any independent random variables $X_1, \ldots, X_n : \Omega \to F$ with $\mathbb{E}[X_i] = 0$, $i = 1, \ldots, n$ we define the filtration $(\mathcal{F}_k)_{k=0,\ldots,n}$ by

$$\mathcal{F}_0 = \{\emptyset, \Omega\} \quad \text{and} \quad \mathcal{F}_k = \sigma(X_1, \ldots, X_k), \quad k = 1, \ldots, n$$

and the $(\mathcal{F}_k)$-martingale $(M_k)_{k=0,\ldots,n}$ by

$$M_0 = 0 \quad \text{and} \quad M_k = \sum_{i=1}^{k} X_i, \quad k = 1, \ldots, n.$$

Then, using (3.5.14) we obtain

$$\mathbb{E}\left[\left\|\sum_{i=1}^{n} X_i\right\|^2\right] = \mathbb{E}[\|M_n\|^2] \leq K \sum_{i=1}^{n} \mathbb{E}[\|M_i - M_{i-1}\|^2] = K \sum_{i=1}^{n} \mathbb{E}[\|X_i\|^2],$$

showing (3.1.13).

Remark 3.5.11 Note that every separable Hilbert space H is a Banach space of M-type 2. Indeed, for every H-valued $(\mathcal{F}_k)$-martingale $(M_i)_{i=0,\ldots,n}$ with $M_0 = 0$ we obtain

$$\mathbb{E}[\|M_n\|^2] = \mathbb{E}\left[\left\|\sum_{i=1}^{n}(M_i - M_{i-1})\right\|^2\right]$$

$$= \mathbb{E}\left[\left\langle \sum_{i=1}^{n}(M_i - M_{i-1}), \sum_{i=1}^{n}(M_i - M_{i-1})\right\rangle\right]$$

$$= \sum_{i=1}^{n} \mathbb{E}[\|M_i - M_{i-1}\|^2] + 2 \sum_{i<j} \mathbb{E}[\langle M_i - M_{i-1}, M_j - M_{j-1}\rangle].$$

For $i < j$ we get, by using that $M_i - M_{i-1}$ is $\mathcal{F}_i$-measurable and that M is a martingale,

$$\mathbb{E}[\langle M_i - M_{i-1}, M_j - M_{j-1}\rangle] = \mathbb{E}[\mathbb{E}[\langle M_i - M_{i-1}, M_j - M_{j-1}\rangle \mid \mathcal{F}_i]]$$
$$= \mathbb{E}[\langle M_i - M_{i-1}, \mathbb{E}[M_j - M_{j-1} \mid \mathcal{F}_i]\rangle] = 0,$$

and hence we arrive at

$$\mathbb{E}[\|M_n\|^2] = \sum_{i=1}^{n} \mathbb{E}[\|M_i - M_{i-1}\|^2],$$

showing that (3.5.14) is fulfilled.

Proposition 3.5.12 *Suppose that F is a Banach space of M-type 2. Then we have*

$$\mathbb{E}\left[\left\| \int_0^T \int_E f(s,x)q(ds,dx) \right\|^2\right]$$

$$\leq K^2 \mathbb{E}\left[\int_0^T \int_E \|f(s,x)\|^2 \beta(dx)ds\right] \quad \textit{for all } f \in \Sigma(F), \tag{3.5.15}$$

where the constant $K > 0$ stems from (3.5.14).

Proof Let $f \in \Sigma(F)$ be a simple function of the form (A.1.1). Define the random variables $(M_i)_{i=0,\ldots,n}$ by

$$M_i = \sum_{k=1}^{i} \sum_{l=1}^{m} a_{k,l} \mathbb{1}_{F_{k,l}} q((t_{k-1}, t_k] \times A_{k,l})$$

$$= \int_0^{t_i} \int_E f(s,x)q(ds,dx), \quad i = 0, \ldots, n.$$

Then we have $M_0 = 0$ and $(M_i)_{i=0,\ldots,n}$ is a $(\mathcal{G}_i)$-martingale, where $(\mathcal{G}_i)_{i=0,\ldots,n}$ denotes the filtration

$$\mathcal{G}_i = \mathcal{F}_{t_i}, \quad i = 0, \ldots, n.$$

For fixed $k = 1, \ldots, n$ we define the random variables $(M_{k,j})_{j=0,\ldots,m}$ by

$$M_{k,j} = \sum_{l=1}^{j} a_{kl} \mathbb{1}_{F_{k,l}} q((t_{k-1}, t_k] \times A_{kl}), \quad j = 0, \ldots, m$$

and the filtration $(\mathcal{G}_{k,j})_{j=0,\ldots,m}$ by

$$\mathcal{G}_{k,j} = \sigma(q((t_{k-1}, t_k] \times A_{k,j})), \quad j = 0, \ldots, m.$$

Then we have $M_{k,0} = 0$ and $(M_{k,j})_{j=0,\ldots,m}$ is a $(\mathcal{G}_{k,j})_{j=0,\ldots,m}$-martingale. Indeed, let $j_1 < j_2$ be arbitrary. Then for $l = j_1 + 1, \ldots, j_2$ the set $F_{k,l} \in \mathcal{F}_{t_{k-1}}$ and $\mathcal{G}_{k,j_1}$ by Definition 2.4.5. Since the product sets $(t_{k-1}, t_k] \times A_{k,l}$ and $(t_{k-1}, t_k] \times A_{k,j_1}$ are disjoint, the random variable $q((t_{k-1}, t_k] \times A_{k,l})$ and $\mathcal{G}_{k,j_1}$ are independent by Theorem 2.4.6. Moreover, the set $F_{k,l}$ and $q((t_{k-1}, t_k] \times A_{k,l})$ are independent. Therefore, we obtain

$$
\begin{aligned}
\mathbb{E}[M_{k,j_2} - M_{k,j_1} \mid \mathcal{G}_{k,j_1}] &= \sum_{l=j_1+1}^{j_2} \mathbb{E}[\mathbb{1}_{F_{k,l}} q((t_{k-1}, t_k] \times A_{k,l}) \mid \mathcal{G}_{k,j_1}] \\
&= \sum_{l=j_1+1}^{j_2} \mathbb{E}[\mathbb{1}_{F_{k,l}} q((t_{k-1}, t_k] \times A_{k,l})] \\
&= \sum_{l=j_1+1}^{j_2} \mathbb{P}(F_{k,l}) \mathbb{E}[q((t_{k-1}, t_k] \times A_{k,l})] = 0,
\end{aligned}
$$

showing that $(M_{k,j})_{j=0,\ldots,m}$ is a $(\mathcal{G}_{k,j})_{j=0,\ldots,m}$-martingale. Therefore, using the estimate (3.5.14), we obtain

$$
\begin{aligned}
&\mathbb{E}\left[\left\| \int_0^T \int_E f(s,x) q(ds, dx) \right\|^2\right] \\
&= \mathbb{E}\left[\left\| \sum_{k=1}^{n} \sum_{l=1}^{m} a_{k,l} \mathbb{1}_{F_{k,l}} q((t_k, t_{k+1}] \times A_{k,l}) \right\|^2\right] \\
&= \mathbb{E}[\|M_n\|^2] \le K \sum_{k=1}^{n} \mathbb{E}[\|M_k - M_{k-1}\|^2] \\
&= K \sum_{k=1}^{n} \mathbb{E}\left[\left\| \sum_{l=1}^{m} a_{k,l} \mathbb{1}_{F_{k,l}} q((t_{k-1}, t_k] \times A_{k,l}) \right\|^2\right] \\
&= K \sum_{k=1}^{n} \mathbb{E}[\|M_{k,m}\|^2] \le K^2 \sum_{k=1}^{n} \sum_{l=1}^{m} \mathbb{E}[\|M_{k,l} - M_{k,l-1}\|^2] \\
&= K^2 \sum_{k=1}^{n} \sum_{l=1}^{m} \mathbb{E}\left[\left\| a_{k,l} \mathbb{1}_{F_{k,l}} q((t_k, t_{k+1}] \times A_{k,l}) \right\|^2\right] \\
&= K^2 \sum_{k=1}^{n} \sum_{l=1}^{m} \|a_{k,l}\|^2 \mathbb{P}(F_{k,l})(t_{k+1} - t_k) \beta(A_{k,l})
\end{aligned}
$$

$$= K^2 \int_0^T \int_E \mathbb{E}[\|f(t,x)\|^2]\beta(dx)dt,$$

proving (3.5.15). $\qquad\qquad\qquad\qquad\qquad\qquad\qquad\qquad\qquad\qquad\qquad$ □

Consequently, if the Banach space F is of M-type 2, then we can define the Itô integral for all $f \in \mathcal{L}^2_{T,\beta}(F)$ as the continuous linear operator

$$\mathcal{L}^2_{T,\beta}(F) \to \mathcal{M}^2_T(F), \quad f \mapsto \left(\int_0^t \int_E f(s,x)q(ds,dx) \right)_{t \in [0,T]}, \qquad (3.5.16)$$

which is the unique extension of (3.5.4), and we obtain the estimate

$$\mathbb{E}\left[\left\| \int_0^T \int_E f(s,x)q(ds,dx) \right\|^2 \right]$$

$$\leq K^2 \mathbb{E}\left[\int_0^T \int_E \|f(s,x)\|^2 \beta(dx)ds \right] \quad \text{for all } f \in \mathcal{L}^2_{T,\beta}(F), \qquad (3.5.17)$$

where the constant $K > 0$ stems from (3.5.14).

Remark 3.5.13 Let $(\Omega, \mathcal{F}, \mu)$ be a measure space with σ-finite measure μ, and $2 \leq p < \infty$, then $F = L^p(\Omega, \mathcal{F}, \mu; \mathbb{R})$ is a Banach space of M-type 2 [85, 100]. For an example of type 2 spaces which is not of M-type 2 we refer to the forthcoming book on Martingales in Banach spaces by G.Pisier

There are separable Banach spaces which are not of M-type 2 (e.g., see Example 3.5.15), but where inequality (3.5.7) is satisfied for certain Poisson random measures. In particular, the following proposition (from [68]) holds for any separable Banach space F.

Proposition 3.5.14 *Suppose that* $\beta(E) < \infty$. *Then inequality* (3.5.9) *is satisfied with* $K_\beta = 4 + 6T\beta(E)$.

Proof Let $f \in \Sigma_T(E/F)$ be arbitrary. Then we have

$$\mathbb{E}\left[\left\| \int_0^T \int_E f(s,x)q(ds,dx) \right\|^2 \right]$$

$$\leq 2\mathbb{E}\left[\left(\int_0^T \int_E \|f(s,x)\| N(ds,dx) \right)^2 \right] + 2\mathbb{E}\left[\left(\int_0^T \int_E \|f(s,x)\| \beta(dx)ds \right)^2 \right]$$

$$\leq 2\mathbb{E}\left[\left(\int_0^T\int_E \|f(s,x)\|q(ds,dx) + \int_0^T\int_E \|f(s,x)\|\beta(dx)ds\right)^2\right]$$

$$+ 2\mathbb{E}\left[\left(\int_0^T\int_E \|f(s,x)\|\beta(dx)ds\right)^2\right].$$

Thus, by the Itô isometry for real-valued integrands and the Cauchy–Schwarz inequality we obtain

$$\mathbb{E}\left[\left\|\int_0^T\int_E f(s,x)q(ds,dx)\right\|^2\right]$$

$$\leq 4\mathbb{E}\left[\left(\int_0^T\int_E \|f(s,x)\|q(ds,dx)\right)^2\right] + 6\mathbb{E}\left[\left(\int_0^T\int_E \|f(s,x)\|\beta(dx)ds\right)^2\right]$$

(3.5.18)

$$\leq 4\mathbb{E}\left[\int_0^T\int_E \|f(s,x)\|^2\beta(dx)ds\right] + 6T\beta(E)\mathbb{E}\left[\int_0^T\int_E \|f(s,x)\|^2\beta(dx)ds\right].$$

(3.5.19)

Consequently, inequality (3.5.9) is satisfied with $K_\beta = 4 + 6T\beta(E)$.   □

Here we provide an example of a separable Banach space which is not of M-type 2 [68].

Example 3.5.15 Let l^1 be the space of all real-valued sequences $(x_j)_{j\in\mathbb{N}} \subset \mathbb{R}$ which are absolutely convergent, that is

$$\|x\|_{l^1} := \sum_{j=1}^\infty |x_j| < \infty.$$

Then $(l^1, \|\cdot\|_{l^1})$ is a separable Banach space which is not M-type 2. Indeed, let $(e_j)_{j\in\mathbb{N}}$ be the standard unit sequences in l^1, which are given by

$$e_1 = (1,0,\ldots), \quad e_2 = (0,1,0\ldots),\ldots$$

Let $n \in \mathbb{N}$ be arbitrary. We denote by $(X_j^{(n)})_{j=1,\ldots,n}$ independent random variables having a normal distribution $N(0,1/n)$, and we define the l^1-valued process $M^{(n)} = (M_k^{(n)})_{k=0,\ldots,n}$ as

$$M_0^{(n)} := 0 \quad \text{and} \quad M_k^{(n)} := \sum_{j=1}^{k} X_j^{(n)} e_j, \quad k = 1, \ldots, n.$$

Then $M^{(n)}$ is a martingale with respect to the filtration $(\mathcal{F}_k^{(n)})_{k=0,\ldots,n}$ given by

$$\mathcal{F}_0^{(n)} = \{\emptyset, \Omega\} \quad \mathcal{F}_k^{(n)} = \sigma(X_1^{(n)}, \ldots, X_k^{(n)}), \quad k = 1, \ldots, n.$$

Moreover, we have

$$\sum_{k=1}^{n} \mathbb{E}[\|M_k^{(n)} - M_{k-1}^{(n)}\|_{l^1}^2] = \sum_{k=1}^{n} \mathbb{E}[\|X_k^{(n)} e_k\|_{l^1}^2] = \sum_{k=1}^{n} \mathbb{E}[|X_k^{(n)}|^2] = 1$$

as well as

$$\mathbb{E}[\|M_n^{(n)}\|_{l^1}^2] = \mathbb{E}[\|\sum_{j=1}^{n} X_j^{(n)} e_j\|_{l^1}^2] = \mathbb{E}\left[\left(\sum_{j=1}^{n} |X_j^{(n)}|\right)^2\right]$$

$$= \sum_{i=1}^{n} \sum_{j=1}^{n} \mathbb{E}[|X_i^{(n)}||X_j^{(n)}|]$$

$$= \sum_{j=1}^{n} \mathbb{E}[|X_i^{(n)}|^2] + \sum_{i=1}^{n} \sum_{j \neq i}^{n} \mathbb{E}[|X_i^{(n)}||X_j^{(n)}|]$$

$$= 1 + \sum_{j \neq i}^{n} \frac{2}{\pi n} = 1 + \frac{2n(n-1)}{\pi n}$$

$$= 1 + \frac{2(n-1)}{\pi} \to \infty \quad \text{for} \quad n \to \infty.$$

Similarly to Proposition 2.1.5 we prove the following result:

Proposition 3.5.16 *Let F, G be Banach spaces of M-type 2, let $A : D(A) \subset F \to G$ be a closed operator and let $f \in \mathcal{L}_{T,\beta}^2(D(A))$ be a function. Then we have $f \in \mathcal{L}_{T,\beta}^2(F)$, $Af \in \mathcal{L}_{T,\beta}^2(G)$ and*

$$A \int_0^t \int_E f(s,x) q(ds,dx) = \int_0^t \int_E Af(s,x) q(ds,dx), \quad t \in [0,T].$$

It will be useful to extend the Itô integral (3.5.16) further. Let $\mathcal{L}_{\infty,\beta}^2(F)$ be the linear space of all progressively measurable functions $f : \Omega \times \mathbb{R}_+ \times E \to F$ such that $f|_{\Omega \times [0,T] \times E} \in \mathcal{L}_{T,\beta}^2(F)$ for all $T > 0$. For all $f \in \mathcal{L}_{\infty,\beta}^2(F)$ we can define the Itô integral

$$\left(\int_0^t \int_E f(s,x) q(ds,dx) \right)_{t \geq 0},$$

which is again a martingale.

For fixed $T > 0$ let $\mathcal{K}^2_{T,\beta}(F) = \mathcal{K}^2_{T,\beta}(E/F)$ be the linear space of all progressively measurable functions $f : \Omega \times [0, T] \times E \to F$ such that

$$\mathbb{P}\left(\int_0^T \int_E \|f(s, x)\|^2 \beta(dx)ds < \infty\right) = 1.$$

For $f \in \mathcal{K}^2_{T,\beta}(F)$ we define the sequence of stopping times

$$\tau_n := \inf\left\{t \in [0, T] : \int_0^t \int_E \|f(s, x)\|^2 \beta(dx)ds \geq n\right\}, \quad n \in \mathbb{N}.$$

Note that $f \mathbb{1}_{[0,\tau_n]} \in \mathcal{L}^2_{T,\beta}(F)$ for all $n \in \mathbb{N}$. Hence, we can define the Itô integral

$$\int_0^t \int_E f(s, x)q(ds, dx) := \lim_{n\to\infty} \int_0^t \int_E f(s, x)\mathbb{1}_{[0,\tau_n]}q(ds, dx), \quad t \in [0, T]$$

which is a local martingale.

Finally, let $\mathcal{K}^2_{\infty,\beta}(F) = \mathcal{K}^2_{\infty,\beta}(E/F)$ be the linear space of all progressively measurable functions $f : \Omega \times \mathbb{R}_+ \times E \to F$ such that $f|_{\Omega \times [0,T] \times E} \in \mathcal{K}^2_{T,\beta}(F)$ for all $T > 0$. For all $f \in \mathcal{K}^2_{\infty,\beta}(F)$ we can define the Itô integral

$$\left(\int_0^t \int_E f(s, x)q(ds, dx)\right)_{t\geq 0},$$

which is again a local martingale.

In the sequel we will use the following result from [93]:

Theorem 3.5.17 *Let $f \in \mathcal{K}^2_{T,\beta}(E/F)$ be arbitrary and let $(f_n)_{n\in\mathbb{N}}$ be a sequence such that $f_n \in \mathcal{K}^2_{T,\beta}(E/F)$ for all $n \in \mathbb{N}$. Suppose that f_n converges $\nu \otimes \mathbb{P}$-almost surely to f on $\Omega \times [0, T] \times E$, when $n \to \infty$, and $\mathbb{P}$-almost surely*

$$\lim_{n\to\infty} \int_0^T \int_E \|f_n - f\|^2 d\nu = 0.$$

Assume there is a $g \in \mathcal{K}^2_{T,\beta}(E/F)$ such that

$$\int_0^T \int_E \|f_n\|^2 d\nu \leq \int_0^T \int_E \|g\|^2 d\nu.$$

Then we have

$$\int_0^t \int_E f(s,x)q(ds,dx) = \lim_{n\to\infty} \int_0^t \int_E f_n(s,x)q(ds,dx),$$

where the limit is in probability.

Proof The proof follows from Theorem 7.7 and Remark 7.8 in [93]. □

For the rest of this section, let F be a separable Banach space. We have a linear operator

$$\Sigma_T(F) \to \mathcal{M}_T^1(F), \quad f \mapsto \left(\int_0^t \int_E f(s,x)q(ds,dx)\right)_{t\in[0,T]}$$

which is continuous according to Lemma 2.4.15. Thus, and because $\Sigma_T(F)$ is dense in $\mathcal{L}_{T,\beta}^1(F)$, we can define the Itô integral (on $\mathcal{L}_{T,\beta}^1(F)$) as the unique extension

$$\mathcal{L}_{T,\beta}^1(F) \to \mathcal{M}_T^1(F), \quad f \mapsto \left(\int_0^t \int_E f(s,x)q(ds,dx)\right)_{t\in[0,T]}. \qquad (3.5.20)$$

Remark 3.5.18 If the separable Banach space F is of M-type 2, then the two integral operators (3.5.16) and (3.5.20) coincide on $\mathcal{L}_{T,\beta}^1(F) \cap \mathcal{L}_{T,\beta}^2(F)$, due to their continuity.

Lemma 3.5.19 *For all $f \in \mathcal{L}_\beta^1(\mathcal{P}_T; F) = L_{T,\text{pred}}^1(F)$ we have*

$$\int_0^t \int_E f(s,x)q(ds,dx) = \int_0^t \int_E f(s,x)N(ds,dx)$$

$$-\int_0^t \int_E f(s,x)\beta(dx)ds, \quad t \in [0,T]. \qquad (3.5.21)$$

Proof For every simple function $f \in \Sigma_T(F)$ identity (3.5.21) holds true by inspection. By Proposition 2.1.6 and Theorem 3.4.2, the linear space $\Sigma_T(F)$ is dense in $\mathcal{L}_\beta^1(F)$. The continuity of the integral operator (2.4.8) yields that (3.5.21) is valid for all $f \in \mathcal{L}_\beta^1(\mathcal{P}_T; F)$. □

Lemma 3.5.20 *For each* $f \in \mathcal{L}^1_{T,\beta}(F) = L^1_{T,\mathrm{ad}}(F)$ *we have*

$$\mathbb{E}\left[\left\|\int_0^t \int_E f(s,x)q(ds,dx)\right\|\right] \le 2\mathbb{E}\left[\int_0^t \int_E \|f(s,x)\|\beta(dx)ds\right], \quad t \in [0,T].$$

Proof According to Theorem 3.4.2 there exists a $g \in L^1_{T,\mathrm{pred}}(F)$ such that $f = g$ almost everywhere with respect to $\mathbb{P} \otimes \beta \otimes \lambda$. Using Lemmas 3.5.19 and 2.4.14, for each $t \in [0,T]$ we obtain

$$\mathbb{E}\left[\left\|\int_0^t \int_E f(s,x)q(ds,dx)\right\|\right] = \mathbb{E}\left[\left\|\int_0^t \int_E g(s,x)q(ds,dx)\right\|\right]$$

$$\le \mathbb{E}\left[\left\|\int_0^t \int_E g(s,x)N(ds,dx)\right\|\right]$$

$$+ \mathbb{E}\left[\left\|\int_0^t \int_E g(s,x)\beta(ds,dx)\right\|\right]$$

$$\le 2\mathbb{E}\left[\int_0^t \int_E \|g(s,x)\|\beta(dx)ds\right]$$

$$= 2\mathbb{E}\left[\int_0^t \int_E \|f(s,x)\|\beta(dx)ds\right],$$

completing the proof.   □

3.6 Integration with Respect to Martingales

Let F be a Banach space of M-type 2. As shown in the previous Sect. 3.5, for $T > 0$ and $f \in \mathcal{L}^2_{T,\beta}(F)$ we can define the Itô integral

$$M_t = \int_0^t \int_E f(s,x)q(ds,dx), \quad t \in [0,T]$$

and $M = (M_t)_{t \in [0,T]}$ is an F-valued martingale. In [86] the stochastic integral with respect to a martingale has been defined. In this section, we shall examine the connection of the integral in [86] to the Itô integral of the previous Sect. 3.5.

Let F be a separable Banach space and fix $T > 0$.

Definition 3.6.1 A martingale $M \in \mathcal{M}_T^2(F)$ is *controlled* by a non-decreasing, real-valued, absolutely continuous process $A = (A_t)_{t\in[0,T]}$ if we have

$$\mathbb{E}[\|M_t - M_s\|^2 \mid \mathcal{F}_s] \le \mathbb{E}[A_t - A_s \mid \mathcal{F}_s], \quad 0 \le s \le t \le T.$$

Lemma 3.6.2 *Suppose the Banach space F is of M-type 2. Then for each $f \in \mathcal{L}_{T,\beta}^2(F)$ the martingale*

$$M_t = \int\limits_0^t \int\limits_E f(s,x)q(ds,dx), \quad t \in [0,T] \qquad (3.6.1)$$

is controlled by the process

$$A_t = K^2 \int\limits_0^t \int\limits_E \|f(s,x)\|^2 \beta(dx)ds, \quad t \in [0,T], \qquad (3.6.2)$$

where the constant $K > 0$ stems from (3.5.14).

Proof Let $0 \le s \le t \le T$ be arbitrary. It suffices to show that for all $B \in \mathcal{F}_s$ we have

$$\mathbb{E}[\|M_t - M_s\|^2 \mathbb{1}_B] \le \mathbb{E}[(A_t - A_s)\mathbb{1}_B].$$

Note that the function $g : \Omega \times [0,T] \times E \to F$ given by

$$g(u,x) = f(u,x)\mathbb{1}_{(s,t]}(v)\mathbb{1}_B$$

is again progressively measurable, and hence $g \in \mathcal{L}_{T,\beta}^2(F)$. Using the estimate (3.5.17), we obtain

$$\mathbb{E}[\|M_t - M_s\|^2 \mathbb{1}_B] = \mathbb{E}\left[\left\| \int\limits_0^T \int\limits_E g(u,x)q(du,dx) \right\|^2\right]$$

$$\le K^2 \mathbb{E}\left[\int\limits_0^T \int\limits_E \|g(u,x)\|^2 \beta(dx)du \right] = \mathbb{E}[(A_t - A_s)\mathbb{1}_B],$$

finishing the proof. □

Now let G be another separable Banach space. We set $\Sigma_T(F,G) = \Sigma(\tilde{\mathcal{G}}_T; L(F,G))$. Note that any process $H \in \Sigma_T(F,G)$ is of the form

$$H = \sum_{k=1}^{n} \Phi_k \mathbb{1}_{F_k} \mathbb{1}_{(t_{k-1}, t_k]} \qquad (3.6.3)$$

for $n \in \mathbb{N}$ with:

- continuous linear operators $\Phi_k \in L(F, G)$ for $k = 1, \ldots, n$;
- time points $0 \leq t_0 \leq \ldots \leq t_n \leq T$;
- sets $F_{k,l} \in \mathcal{F}_{t_{k-1}}$ for $k = 1, \ldots, n$.

For a simple process $H \in \Sigma_T(F, G)$ of the form (3.6.3) and a square-integrable martingale $M \in \mathcal{M}_T^2(F)$ we define the *Itô integral*

$$(H \cdot M)_t := \int_0^t H_s dM_s := \sum_{k=1}^{n} \mathbb{1}_{F_k} \Phi_k (M_{t_k \wedge t} - M_{t_{k-1} \wedge t}), \quad t \in [0, T]. \qquad (3.6.4)$$

Lemma 3.6.3 *For each $H \in \Sigma_T(F, G)$ and $M \in \mathcal{M}_T^2(F)$ we have $H \cdot M \in \mathcal{M}_T^2(G)$.*

Proof Let $H \in \Sigma_T(F, G)$ be an arbitrary process of the form (3.6.3). Let $k = 1, \ldots, n$ be arbitrary. Then, the process

$$N_t = \Phi_k (M_{t_k \wedge t} - M_{t_{k-1} \wedge t}), \quad t \in [0, T]$$

is a martingale. We shall now prove that $\mathbb{1}_{F_k} N$ is also a martingale, which will finish the proof, as the Itô integral is given by (3.6.4). Let $0 \leq s \leq t \leq T$ be arbitrary. If $s \leq t_{k-1}$, then we have

$$\mathbb{E}[\mathbb{1}_{F_k} N_t \mid \mathcal{F}_s] = \mathbb{E}[\mathbb{E}[\mathbb{1}_{F_k} N_t \mid \mathcal{F}_{t_{k-1}}] \mid \mathcal{F}_s] = \mathbb{E}[\mathbb{1}_{F_k} \mathbb{E}[N_t] \mid \mathcal{F}_s] = 0 = \mathbb{1}_{F_k} N_s,$$

and for $s > t_k$ we obtain

$$\mathbb{E}[\mathbb{1}_{F_k} N_t \mid \mathcal{F}_s] = \mathbb{1}_{F_k} \mathbb{E}[N_t \mid \mathcal{F}_s] \mathbb{1}_{F_k} N_s,$$

showing that $\mathbb{1}_{F_k} N$ is a martingale. □

Due to Lemma 3.6.3, for any $M \in \mathcal{M}_T^2(F)$ we may regard the Itô integral defined in (3.6.4) as a linear operator

$$\Sigma_T(F, G) \to \mathcal{M}_T^2(G), \quad H \mapsto H \cdot M. \qquad (3.6.5)$$

Note that for any $H \in \Sigma_T(F, G)$ the mapping

$$\mathcal{M}_T(F) \to \mathcal{M}_T(G), \quad M \mapsto H \cdot M$$

is also well-defined and linear.

Proposition 3.6.4 *Let $M \in \mathcal{M}_T^2(F)$ be arbitrary. Suppose the Banach space G is of M-type 2 and that there exists a process $A = (A_t)_{t \in [0,T]}$ such that M is controlled by A. Then we have*

$$\mathbb{E}\left[\left\|\int_0^T H_s dM_s\right\|^2\right] \le K_G \mathbb{E}\left[\int_0^T \|H_s\|^2 dA_s\right] \quad \text{for all } f \in \Sigma_T(F, G), \qquad (3.6.6)$$

where the constant $K_G > 0$ stems from estimate (3.5.14) regarding the Banach space G.

Proof Let $f \in \Sigma_T(F, G)$ be a simple function of the form (3.6.3). Define the random variables $(N_i)_{i=0,\dots,n}$ by

$$N_i = \sum_{k=1}^{i} \mathbb{1}_{A_k} \Phi_k (M_{t_k} - M_{t_{k-1}}) = (H \cdot M)_{t_i}, \quad i = 0, \dots, n.$$

Then we have $N_0 = 0$ and $(N_i)_{i=0,\dots,n}$ is a $(\mathcal{G}_i)$-martingale, where $(\mathcal{G}_i)_{i=0,\dots,n}$ denotes the filtration

$$\mathcal{G}_i = \mathcal{F}_{t_i}, \quad i = 0, \dots, n.$$

Hence, using the estimate (3.5.14), we obtain

$$\mathbb{E}\left[\left\|\int_0^T H_s dM_s\right\|^2\right] = \mathbb{E}[\|N_n\|^2] \le K_G \sum_{k=1}^{n} \mathbb{E}[\|N_k - N_{k-1}\|^2]$$

$$= K_G \sum_{k=1}^{n} \mathbb{E}[\|\mathbb{1}_{F_k} \Phi_k (M_{t_k} - M_{t_{k-1}})\|^2]$$

$$\le K_G \sum_{k=1}^{n} \mathbb{E}[\mathbb{1}_{F_k} \|\Phi_k\|^2 \|M_{t_k} - M_{t_{k-1}}\|^2]$$

$$= K_G \sum_{k=1}^{n} \mathbb{E}[\mathbb{1}_{F_k} \|\Phi_k\|^2 \mathbb{E}[\|M_{t_k} - M_{t_{k-1}}\|^2 \mid \mathcal{F}_{t_{k-1}}]]$$

$$\le K_G \sum_{k=1}^{n} \mathbb{E}[\mathbb{1}_{F_k} \|\Phi_k\|^2 \mathbb{E}[A_{t_k} - A_{t_{k-1}} \mid \mathcal{F}_{t_{k-1}}]]$$

$$= K_G \mathbb{E}\left[\sum_{k=1}^{n} \mathbb{1}_{F_k} \|\Phi_k\|^2 (A_{t_k} - A_{t_{k-1}})\right] = K_G \mathbb{E}\left[\int_0^T \|H_s\|^2 dA_s\right],$$

completing the proof. $\square$

For a non-decreasing, real-valued, absolutely continuous process $A = (A_t)_{t \in [0,T]}$ we set $\mu = \mathbb{P} \otimes A$ and $\mathcal{L}^2_{T,A}(F, G) = L^2_{T,\mathrm{ad}}(L(F, G))$ in the sense of definition (3.4.1) with $\tilde{\Omega} = \Omega$. By Proposition 2.1.6 and Theorem 3.4.2, the linear space $\Sigma_T(F, G)$ is dense in $\mathcal{L}^2_{T,A}(F, G)$.

Let $M \in \mathcal{M}^2_T(F)$ be a square-integrable martingale. Suppose that the Banach space G is of M-type 2 and that there exists a process $A = (A_t)_{t \in [0,T]}$ such that M is controlled by A. According to Lemma 3.6.2, the latter condition is in particular satisfied if the Banach space F is of M-type 2 and martingale M is given by (3.6.1) for some $f \in \mathcal{L}^2_{t,\beta}(F)$. By Proposition 3.6.4, the integral operator (3.6.5) is continuous. Since $\Sigma_T(F, G)$ is dense in $\mathcal{L}^2_{T,A}(F, G)$, it has a unique extension

$$\mathcal{L}^2_{T,A}(F, G) \to \mathcal{M}^2_T(G), \quad H \mapsto H \cdot M, \tag{3.6.7}$$

which we also call the *Itô integral*, and we have the estimate

$$\mathbb{E}\left[\left\| \int_0^T H_s dM_s \right\|^2 \right] \leq K_G \mathbb{E}\left[\int_0^T \|H_s\|^2 dA_s \right] \quad \text{for all } f \in \mathcal{L}^2_{T,A}(F, G). \tag{3.6.8}$$

Theorem 3.6.5 *Let F, G be Banach spaces of M-type 2 and let $f \in \mathcal{L}^2_{T,\beta}(F)$ be arbitrary. Let $M \in \mathcal{M}^2_T(F)$ and $A = (A_t)_{t \in [0,T]}$ be given by (3.6.1) and (3.6.2). Then, for all $H \in \mathcal{L}^2_{T,A}(F, G)$, we have $Hf \in \mathcal{L}^2_{T,\beta}(G)$ and*

$$\left(\int_0^t H_s dM_s \right)_{t \in [0,T]} = \left(\int_0^t \int_E H_s f(s, x) q(ds, dx) \right)_{t \in [0,T]}. \tag{3.6.9}$$

Proof For all $H \in \mathcal{L}^2_{T,A}(F, G)$ we have

$$\mathbb{E}\left[\int_0^T \int_E \|H_s f(s, x)\|^2 \beta(dx) ds \right] \leq \mathbb{E}\left[\int_0^T \int_E \|H_s\|^2 \|f(s, x)\|^2 \beta(dx) ds \right]$$

$$= \mathbb{E}\left[\int_0^T \|H_s\|^2 dA_s \right] < \infty,$$

showing that $Hf \in \mathcal{L}^2_{T,\beta}(G)$. The proof of identity (3.6.9) is divided into several steps:

1. For elementary integrands $f \in \Sigma_T(F)$ and $H \in \Sigma_T(F, G)$ identity (3.6.9) holds true by inspection.

2. Now let $f \in \mathcal{L}^2_{T,\beta}(F)$ and $H \in \Sigma_T(F, G)$ be arbitrary. By Proposition 2.1.6 and Theorem 3.4.2 there exists a sequence $(f_n)_{n \in \mathbb{N}} \subset \Sigma_T(F)$ of simple functions such that $f_n \to f$ in $\mathcal{L}^2_{T,\beta}(F)$. For each $n \in \mathbb{N}$ let $M^n \in \mathcal{M}^2_T(F)$ be defined by

$$M^n_t = \int_0^t \int_E f_n(s, x) q(ds, dx), \quad t \in [0, T].$$

Moreover, we define $N^n \in \mathcal{M}^2_T(F)$ by

$$N^n_t = \int_0^t \int_E (f(s, x) - f_n(s, x)) q(ds, dx), \quad t \in [0, T]$$

and the process $B^n = (B^n_t)_{t \in [0,T]}$ by

$$B^n_t = \int_0^t \int_E \|f(s, x) - f_n(s, x)\|^2 \beta(dx) ds, \quad t \in [0, T].$$

By Lemma 3.6.2 the martingale N^n is controlled by B^n. Hence, by (3.6.8) we get

$$\mathbb{E}\left[\left\| \int_0^T H_s dM_s - \int_0^T H_s M^n_s \right\|^2\right]$$

$$= \mathbb{E}\left[\left\| \int_0^T H_s dN^n_s \right\|^2\right] \leq K_G \mathbb{E}\left[\int_0^T \|H_s\|^2 dB^n_s \right]$$

$$= K_G \mathbb{E}\left[\int_0^T \int_E \|f(s, x) - f_n(s, x)\|^2 \beta(dx) ds \right] \to 0,$$

showing that

$$H \cdot M^n \to H \cdot M \quad \text{in } \mathcal{M}^2_T(G). \tag{3.6.10}$$

Moreover, we have $Hf_n \to Hf$ in $\mathcal{L}^2_{t,\beta}(F)$, because

$$\mathbb{E}\left[\int_0^t \int_E \|H_s f_n(s, x) - H_s f(s, x)\|^2 \beta(dx) ds \right]$$

$$\leq \mathbb{E}\left[\int\limits_0^t \int\limits_E \|H_s\|^2 \|f_n(s,x) - f(s,x)\|^2 \beta(dx)ds\right]$$

$$\leq \max_{k=1,\ldots,n} \|\Phi_k\|^2 \mathbb{E}\left[\int\limits_0^t \int\limits_E \|f_n(s,x) - f(s,x)\|^2 \beta(dx)ds\right] \to 0.$$

By Step 1, the convergence (3.6.10) and the continuity of the integral operator (3.5.16) we obtain

$$\left(\int\limits_0^t H_s dM_s\right)_{t\in[0,T]} = \lim_{n\to\infty}\left(\int\limits_0^t H_s dM_s^n\right)_{t\in[0,T]}$$

$$= \lim_{n\to\infty}\left(\int\limits_0^t \int\limits_E H_s f_n(s,x)q(ds,dx)\right)_{t\in[0,T]}$$

$$= \left(\int\limits_0^t \int\limits_E H_s f(s,x)q(ds,dx)\right)_{t\in[0,T]},$$

where the limits are taken in $\mathcal{M}_T^2(G)$.

3. Finally, let $f \in \mathcal{L}_{T,\beta}^2(F)$ and $H \in \mathcal{L}_{T,A}^2(F,G)$ be arbitrary. By Proposition 2.1.6 and Theorem 3.4.2 there exists a sequence $(H^n)_{n\in\mathbb{N}} \subset \Sigma_T(F,G)$ of simple processes such that $H^n \to H$ in $\mathcal{L}_{T,A}^2(F,G)$. We have $H^n f \to Hf$ in $\mathcal{L}_{T,\beta}^2(G)$, because

$$\mathbb{E}\left[\int\limits_0^T \int\limits_E \|H_s^n f(s,x) - H_s f(s,x)\|^2 \beta(dx)ds\right]$$

$$\leq \mathbb{E}\left[\int\limits_0^T \int\limits_E \|H_s^n - H_s\|^2 \|f(s,x)\|^2 \beta(dx)ds\right]$$

$$= \mathbb{E}\left[\int\limits_0^T \|H_s^n - H_s\|^2 dA_s\right] \to 0.$$

By Step 2 and the continuity of the integral operators (3.5.16) and (3.6.7) we obtain

$$\left(\int\limits_0^t H_s dM_s\right)_{t\in[0,T]} = \lim_{n\to\infty}\left(\int\limits_0^t H_s^n dM_s\right)_{t\in[0,T]}$$

$$= \lim_{n \to \infty} \left(\int_0^t \int_E H_s^n f(s,x) q(ds,dx) \right)_{t \in [0,T]}$$

$$= \left(\int_0^t \int_E H_s f(s,x) q(ds,dx) \right)_{t \in [0,T]},$$

where the limits are taken in $\mathcal{M}_T^2(G)$. $\square$

3.7 Itô's Formula

We assume that E is a Blackwell space and F is a separable Banach space, $q(dt,dx) = N(dt,dx)(\omega) - \nu(dt,dx)$, with $\nu(dt,dx) = dt \otimes \beta(dx)$ is the compensated Poisson random measure.

Let $0 < t \le T$, $A \in \mathcal{B}(E \backslash \{0\})$ and

$$Z_t := \int_0^t \int_A f(s,x) q(ds,dx).$$

We assume that $f \in \mathcal{K}_{T,\beta}^2(F)$, and that for all $g \in \mathcal{L}_{\infty,\beta}^2(F)$, g is Itô integrable with respect to the compensated Poisson measure $q(ds,dx)$.

In Theorem 3.7.2 we shall prove the Itô formula for the F-valued random process $(Y_t)_{t \ge 0}$, with

$$Y_t := Z_t + \int_0^t \int_A k(s,x) N(ds,dx).$$

We assume that A is a set with $\beta(A) < \infty$, $k : \Omega \times \mathbb{R}_+ \times E \to F$ is a progressively measurable process. Moreover k is càdlàg or càglàd $\beta(dx) \otimes \mathbb{P}$-almost surely.

Improving the result in [93] we do not need to assume here that the Fréchet derivatives of $\mathcal{H}$ are uniformly bounded. Hence important functions $\mathcal{H}$ for applications, as proposed, for instance, in Example 3.7.6 at the end of this section, can be considered for the Itô formula. Instead, we need to introduce the following definition and properties:

Definition 3.7.1 We call a continuous, non-decreasing function $h : \mathbb{R}_+ \to \mathbb{R}_+$ *quasi-sublinear* if there is a constant $C > 0$ such that

$$h(x+y) \le C\big(h(x) + h(y)\big), \quad x,y \in \mathbb{R}_+,$$
$$h(xy) \le Ch(x)h(y), \quad x,y \in \mathbb{R}_+.$$

Theorem 3.7.2 *We suppose that:*

- $\mathcal{H} \in C^{1,2}(\mathbb{R}_+ \times F; G)$ *is a function such that*

$$\|\partial_y \mathcal{H}(s, y)\| \leq h_1(\|y\|), \quad (s, y) \in \mathbb{R}_+ \times F \tag{3.7.1}$$

$$\|\partial_{yy} \mathcal{H}(s, y)\| \leq h_2(\|y\|), \quad (s, y) \in \mathbb{R}_+ \times F \tag{3.7.2}$$

for quasi-sublinear functions $h_1, h_2 : \mathbb{R}_+ \to \mathbb{R}_+$.
- $f : \Omega \times \mathbb{R}_+ \times E \to F$ *is a progressively measurable process such that for all $t \in \mathbb{R}_+$ we have $\mathbb{P}$-almost surely*

$$\int_0^t \int_A \|f(s, x)\|^2 v(ds, dx) + \int_0^t \int_A h_1(\|f(s, x)\|)^2 \|f(s, x)\|^2 v(ds, dx)$$

$$+ \int_0^t \int_A h_2(\|f(s, x)\|) \|f(s, x)\|^2 v(ds, dx) < \infty. \tag{3.7.3}$$

•

$$\int_0^t \int_\Lambda \|k(s, x)\| v(ds, dx) < \infty \quad \mathbb{P}\text{-a.s.}$$

Then the following statements are true:

1. *For all $t \in \mathbb{R}_+$ we have $\mathbb{P}$-almost surely*

$$\int_0^t \|\partial_s \mathcal{H}(s, Y_s)\| ds < \infty, \tag{3.7.4}$$

$$\int_0^t \int_A \|\mathcal{H}(s, Y_s + f(s, x)) - \mathcal{H}(s, Y_s)\|^2 v(ds, dx) < \infty, \tag{3.7.5}$$

$$\int_0^t \int_A \|\mathcal{H}(s, Y_s + f(s, x)) - \mathcal{H}(s, Y_s) - \partial_y \mathcal{H}(s, Y_s) f(s, x)\| v(ds, dx) < \infty,$$

$$\tag{3.7.6}$$

$$\int_0^t \int_\Lambda \|\mathcal{H}(s, Y_{s-} + k(s, x)) - \mathcal{H}(s, Y_{s-})\| N(ds, dx) < \infty. \tag{3.7.7}$$

2. *We have* $\mathbb{P}$-*almost surely*

$$\mathcal{H}(t, Y_t) = \mathcal{H}(0, Y_0) + \int_0^t \partial_s \mathcal{H}(s, Y_s) ds$$

$$+ \int_0^t \int_A \left(\mathcal{H}(s, Y_{s-} + f(s, x)) - \mathcal{H}(s, Y_{s-}) \right) q(ds, dx)$$

$$+ \int_0^t \int_A \left(\mathcal{H}(s, Y_s + f(s, x)) - \mathcal{H}(s, Y_s) \right.$$

$$\left. - \partial_y \mathcal{H}(s, Y_s) f(s, x) \right) v(ds, dx)$$

$$+ \int_0^t \int_\Lambda \left(\mathcal{H}(s, Y_{s-} + k(s, x)) - \mathcal{H}(s, Y_{s-}) \right) N(ds, dx), \quad t \geq 0.$$

$$(3.7.8)$$

Remark 3.7.3 Assume f and k do not depend on $\omega \in \Omega$. Let $\mathcal{L} \in \mathbf{L}(F/\mathbb{R})$ such that

$$\mathcal{L}\mathcal{H}(y) = \int_A \{ \mathcal{H}(y + f(s, x)) - \mathcal{H}(y) - \partial_y \mathcal{H}(y) f(s, x) \} v(ds, dx)$$

$$+ \int_\Lambda \{ \mathcal{H}(y + k(s, x)) - \mathcal{H}(y) \} v(ds, dx). \quad (3.7.9)$$

Then if $\mathcal{H}(s, \cdot) \in Dom(\mathcal{L})$ a.s. for $s \in [0, T]$

$$\mathcal{H}(t, Y_t) - \mathcal{H}(\tau, Y_\tau) = \int_\tau^t \partial_s \mathcal{H}(s, Y_{s-}) ds + \int_\tau^t \mathcal{L}\mathcal{H}(s, Y_{s-})$$

$$+ \int_\tau^t \int_A \{ \mathcal{H}(s, Y_{s-} + f(s, x)) - \mathcal{H}(s, Y_{s-}) \} q(ds, dx)$$

$$+ \int_\tau^t \int_\Lambda \{ \mathcal{H}(s, Y_{s-} + k(s, x)) - \mathcal{H}(s, Y_{s-}) \} q(ds, dx) \quad \mathbb{P}\text{-}a.s.$$

It follows that $\mathcal{L}$ is the generator of $(Y_t)_{t \in [0, T]}$.

Remark 3.7.4 If $k(s, x) = 0$ then $\mathcal{H}(s, \cdot) \in Dom(\mathcal{L})$. If k does not depend on $\omega \in \Omega$ and

$$\int_{\Lambda} \|k(s, x)\|^2 \nu(ds, dx) < \infty$$

and $\partial_y \mathcal{H}(s, y)$ is uniformly bounded, then $\mathcal{H}(s, \cdot) \in Dom(\mathcal{L})$.

The second statement in Remark 3.7.4 is easily checked by proving that the second integral in (3.7.9) is well-defined. This is a consequence of the following inequality, see, e.g., [55, Chap. X], which holds for Fréchét differentiable functions $\mathcal{H} : E \to G$.

$$\|\mathcal{H}(y) - \mathcal{H}(y_0)\| \leq \|y - y_0\| \sup_{0 < \theta \leq 1} \|\mathcal{H}'(y_0 + \theta(y - y_0))\|, \qquad (3.7.10)$$

where $\mathcal{H}'$ denotes the first Fréchet derivative of $\mathcal{H}$.

We shall use besides (3.7.10) the following inequalities for twice Fréchét differentiable functions $\mathcal{H} : E \to G.$, which can be found, for example, in [55, Chap. X],

$$\|\mathcal{H}(y) - \mathcal{H}(y_0) - \mathcal{H}'(y_0)(y - y_0)\|$$
$$\leq \|y - y_0\| \sup_{0 < \theta \leq 1} \|\mathcal{H}'(y_0 + \theta(y - y_0)) - \mathcal{H}'(y_0)\|, \qquad (3.7.11)$$

$$\|\mathcal{H}'(y) - \mathcal{H}'(y_0)\| \leq \|y - y_0\| \sup_{0 < \theta \leq 1} \|\mathcal{H}''(y_0 + \theta(y - y_0))\|. \qquad (3.7.12)$$

Before proving Theorem 3.7.2 we first prove a more restricted result given by the following lemma.

Lemma 3.7.5 *Suppose that:*

- $\mathcal{H} \in C_b^{1,2}(\mathbb{R}_+ \times F; G)$;
- $f : \Omega \times \mathbb{R}_+ \times E \to F$, *and* $k : \Omega \times \mathbb{R}_+ \times E \to F$ *are simple functions.*

Then the following statements are true:

1. *For all* $t \in \mathbb{R}_+$ *we have* $\mathbb{P}$-*almost surely*

$$\int_0^t \|\partial_s H(s, Y_s)\| ds < \infty, \qquad (3.7.13)$$

$$\int_0^t \int_A \|\mathcal{H}(s, Y_s + f(s, x)) - \mathcal{H}(s, Y_s)\| N(ds, dx) < \infty, \qquad (3.7.14)$$

$$\int_0^t \int_A \|\partial_y \mathcal{H}(s, Y_s) f(s, x)\| \nu(ds, dx) < \infty, \qquad (3.7.15)$$

$$\int\limits_0^t \int\limits_\Lambda \|\mathcal{H}(s, Y_{s-} + k(s, x)) - \mathcal{H}(s, Y_{s-})\| N(ds, dx) < \infty. \tag{3.7.16}$$

2. (3.7.8) *holds $\mathbb{P}$-almost surely.*

We will give two proofs of Lemma 3.7.5. One for the case where we make the additional assumption that E is a separable Banach space, and one for the general case where E is Blackwell. In fact assuming that E is a separable Banach space makes the proof very natural as paths are decomposed into pure jump cádlág functions and continuous functions. Once this case is understood the proof for the more general case where E is a Blackwell space appears natural.

Proof We first remark that (3.7.13) and (3.7.15) hold because of the continuity of the partial derivatives $\partial_s H$ and $\partial_y H$ and because $\int_0^t \int_A \|f(s, x)\| \nu(ds, dx) < \infty$, since f is a simple function.

Let us first assume that E is a separable Banach space. In this case, $N(ds, dx)$ is the counting measure of a Lévy process $(X_t)_{t \geq 0}$, and due to Remark 3.5.1

$$Z_t(\omega) = \sum_{0 < s \leq t} f(s, (\Delta X_s)(\omega), \omega) \mathbb{1}_A(\Delta X_s(\omega)) - \int\limits_0^t \int\limits_A f(s, x, \omega) \nu(ds, dx). \tag{3.7.17}$$

Moreover

$$\int\limits_\tau^t \int\limits_A \{\mathcal{H}(s, Y_{s-} + f(s, x)) - \mathcal{H}(s, Y_{s-})\} q(ds, dx)$$

$$= \sum_{\tau < s \leq t} \mathcal{H}(s, Y_{s-}(\omega) + f(s, \Delta X_s(\omega), \omega)) - \mathcal{H}(s, Y_{s-}(\omega)) \mathbb{1}_A(\Delta X_s(\omega))$$

$$- \int\limits_\tau^t \int\limits_A \{\mathcal{H}(s, Y_{s-}(\omega) + f(s, x, \omega)) - \mathcal{H}(s, Y_{s-}(\omega))\} \nu(ds, dx) \quad \mathbb{P}\text{-a.s.}$$

As Lévy processes are continuous in probability, it follows that almost surely

$$\mathbb{1}_\Lambda(\Delta X_s(\omega)) \Delta k(s, \Delta X_s(\omega), \omega) = 0,$$

$$\mathbb{1}_A(\Delta X_s(\omega)) \Delta f(s, \Delta X_s(\omega), \omega) = 0,$$

$$\forall s \in [0, T] \quad \mathbb{1}_A(\Delta X_s(\omega)) \mathbb{1}_\Lambda(\Delta X_s(\omega)) = 0.$$

Let

$$\Gamma_\omega^n(A) := \{k \in 0, \ldots, 2^n - 1 : \exists s \in (\tau_k^n, \tau_{k+1}^n] : \Delta X_s(\omega) \in A\}$$

with $\tau_k^n := \tau + \frac{k(t-\tau)}{2^n}$. Then

$$\mathcal{H}(t, Y_t(\omega)) - \mathcal{H}(\tau, Y_\tau(\omega)) \tag{3.7.18}$$

$$= \sum_{k=0}^{2^n-1} \mathcal{H}(\tau_{k+1}^n, Y_{\tau_{k+1}^n}(\omega)) - \mathcal{H}(\tau_k^n, Y_{\tau_k^n}(\omega))$$

$$= \sum_{k \in \Gamma_\omega^n(A) \cup \Gamma_\omega^n(\Lambda)} \mathcal{H}(\tau_{k+1}^n, Y_{\tau_{k+1}^n}(\omega)) - \mathcal{H}(\tau_k^n, Y_{\tau_k^n}(\omega))$$

$$+ \sum_{k \notin \Gamma_\omega^n(A) \cup \Gamma_\omega^n(\Lambda)} \mathcal{H}(\tau_{k+1}^n, Y_{\tau_{k+1}^n}(\omega)) - \mathcal{H}(\tau_k^n, Y_{\tau_k^n}(\omega)).$$

It can easily be checked that almost surely

$$\lim_{n \to \infty} \sum_{k \in \Gamma_\omega^n(A) \cup \Gamma_\omega^n(\Lambda)} \mathcal{H}(\tau_{k+1}^n, Y_{\tau_{k+1}^n}(\omega)) - \mathcal{H}(\tau_k^n, Y_{\tau_k^n}(\omega)) \tag{3.7.19}$$

$$= \int_\tau^t \int_A \{\mathcal{H}(s, Y_{s_-} + f(s,x)) - \mathcal{H}(s, Y_{s_-})\} q(ds, dx)$$

$$+ \int_\tau^t \int_\Lambda \{\mathcal{H}(s, Y_{s_-} + k(s,x)) - \mathcal{H}(s, Y_{s_-})\} N(ds, dx)$$

$$+ \int_\tau^t \int_A \{\mathcal{H}(s, Y_{s_-} + f(s,x)) - \mathcal{H}(s, Y_{s_-})\} \nu(ds, dx).$$

Equation (3.7.8) follows once we show that for some subsequence of $\{n\}_{n \in \mathbb{N}}$, which for simplicity we still denote by $\{n\}$, the following convergence holds for $n \to \infty$ almost surely

$$\lim_{n \to \infty} \sum_{k \notin \Gamma_\omega^n(A) \cup \Gamma_\omega^n(\Lambda)} \{\mathcal{H}(\tau_{k+1}^n, Y_{\tau_{k+1}^n}(\omega)) - \mathcal{H}(\tau_k^n, Y_{\tau_k^n}(\omega))\}$$

$$= \int_\tau^t \partial_s \mathcal{H}(s, Y_s(\omega)) ds - \int_\tau^t \int_A \partial_y \mathcal{H}(s, Y_{s_-}(\omega)) f(s, x, \omega) \nu(ds, dx). \tag{3.7.20}$$

Proof of (3.7.20): We have the Taylor expansion of the function $\mathcal{H}(s, y)$:

$$\sum_{k \notin \Gamma_\omega^n(A) \cup \Gamma_\omega^n(\Lambda)} (\mathcal{H}(\tau_{k+1}^n, Y_{\tau_{k+1}^n}(\omega)) - \mathcal{H}(\tau_k^n, Y_{\tau_k^n}(\omega)))$$

$$= \sum_{k \notin \Gamma_\omega^n(A) \cup \Gamma_\omega^n(\Lambda)} \partial_y \mathcal{H}(\tau_k^n, Y_{\tau_k^n}(\omega))(Y_{\tau_{k+1}^n}(\omega) - Y_{\tau_k^n}(\omega))$$

$$+ \sum_{k \notin \Gamma_\omega^n(A) \cup \Gamma_\omega^n(\Lambda)} \partial_s \mathcal{H}(\tau_k^n, Y_{\tau_k^n})(\tau_{k+1}^n - \tau_k^n) \qquad (3.7.21)$$

$$+ \sum_{k \notin \Gamma_\omega^n(A) \cup \Gamma_\omega^n(\Lambda)} \text{er}_k^n(\omega).$$

Then we have

$$\lim_{n \to \infty} \sum_{k \notin \Gamma_\omega^n(A) \cup \Gamma_\omega^n(\Lambda)} \partial_s \mathcal{H}(\tau_k^n, Y_{\tau_k^n})(\tau_{k+1}^n - \tau_k^n) = \int_\tau^t \partial_s \mathcal{H}(s, Y_{s_-}(\omega)) ds \quad \mathbb{P}\text{-a.s.}$$

and

$$\lim_{n \to \infty} \sum_{k \notin \Gamma_\omega^n(A) \cup \Gamma_\omega^n(\Lambda)} \partial_y \mathcal{H}(\tau_k^n, Y_{\tau_k^n}(\omega))(Y_{\tau_{k+1}^n}(\omega)) - Y_{\tau_k^n}(\omega))$$

$$= \lim_{n \to \infty} \sum_{k \notin \Gamma_\omega^n(A) \cup \Gamma_\omega^n(\Lambda)} \partial_y \mathcal{H}(\tau_k^n, Y_{\tau_k^n}(\omega))(Z_{\tau_{k+1}^n}(\omega)) - Z_{\tau_k^n}(\omega))$$

$$= - \lim_{n \to \infty} \sum_{k \notin \Gamma_\omega^n(A) \cup \Gamma_\omega^n(\Lambda)} \int_{\tau_k^n}^{\tau_{k+1}^n} \int_A \partial_y \mathcal{H}(s, Y_{s_-}(\omega)) f(s, x, \omega)]$$

$$= - \int_\tau^t \int_A \partial_y \mathcal{H}(s, Y_{s_-}(\omega)) f(s, x, \omega) \nu(ds, dx) \quad \mathbb{P}\text{-a.s.}$$

where the last equality follows because $Y.(\omega)$ is $\mathbb{P}$-a.s. uniformly bounded on $[0, T]$ and from the following estimates:

$$\limsup_{n \to \infty} \sum_{k \notin \Gamma_\omega^n(A) \cup \Gamma_\omega^n(\Lambda)} \left\| \int_{\tau_k^n}^{\tau_{k+1}^n} \int_A (\partial_y \mathcal{H}(\tau_k^n, Y_{\tau_k^n}(\omega)) - \partial_y \mathcal{H}(s, Y_s)(\omega)) \right.$$

$$\left. \times f(s, x, \omega) \nu(ds, dx) \right\|$$

$$\leq C(\omega) \sum_{k \notin \Gamma_\omega^n(A) \cup \Gamma_\omega^n(\Lambda)} \int_{\tau_k^n}^{\tau_{k+1}^n} \int_A \|Y_{\tau_k^n}(\omega) - Y_{s_-}(\omega)\| \|f(s, x, \omega)\| \nu(ds, dx)$$

$$\leq C(\omega) \sup_{k \notin \Gamma_\omega^n(A) \cup \Gamma_\omega^n(\Lambda)} \int_{\tau_k^n}^{\tau_{k+1}^n} \int_A \|f(s, x, \omega)\| \nu(ds, dx).$$

(3.7.20) is proven once we show that in the expansion (3.7.21) we have

$$\lim_{n\to\infty} \sum_{k\notin\Gamma_\omega^n} \mathrm{er}_k^n(\omega) = 0 \quad \mathbb{P}\text{-a.s.}$$

This is a consequence of

$$\|\mathrm{er}_k^n(\omega)\| \le C(\omega)(\|Y_{\tau_{k+1}^n}(\omega) - Y_{\tau_k^n}(\omega)\|^2 + |\tau_{k+1}^n - \tau_k^n|^2, \tag{3.7.22}$$

$$\lim_{n\to\infty} \sum_{k\notin\Gamma_\omega^n(A)\cup\Gamma_\omega^n(\Lambda)} \|Y_{\tau_{k+1}^n}(\omega) - Y_{\tau_k^n}(\omega)\|^2 = 0 \quad \mathbb{P}\text{-a.s.} \tag{3.7.23}$$

$$\sum_{k=0}^{2^n-1} |\tau_{k+1}^n - \tau_k^n|^2 \le \frac{(t-\tau)^2}{2^n}. \tag{3.7.24}$$

Proof of (3.7.22):

$$\|\mathcal{H}(s,y) - \mathcal{H}(s_0,y_0) - \partial_s\mathcal{H}(s_0,y_0)(s-s_0) - \partial_y\mathcal{H}(s_0,y_0)(y-y_0)\|$$

$$\le \|\mathcal{H}(s,y) - \mathcal{H}(s,y_0) - \partial_y\mathcal{H}(s,y_0)(y-y_0)\|$$

$$\quad + \|\mathcal{H}(s,y_0) - \mathcal{H}(s_0,y_0) - \partial_s\mathcal{H}(s_0,y_0)(s-s_0)\|$$

$$\quad + \|(\partial_y\mathcal{H}(s,y_0) - \partial_y\mathcal{H}(s_0,y_0))(y-y_0)\|$$

$$\le \sup_{0<\theta\le1} \|\partial_s\mathcal{H}(s_0+\theta(s-s_0),y_0) - \partial_s\mathcal{H}(s_0,y_0)\||(s-s_0)|$$

$$\quad + \sup_{0<\theta\le1} \|\partial_y\mathcal{H}(s,y_0+\theta(y-y_0)) - \partial_y\mathcal{H}(s,y_0)\|\|(y-y_0)\|$$

$$\quad + \|\partial_y\mathcal{H}(s,y_0)(y-y_0) - \partial_y\mathcal{H}(s_0,y_0)(y-y_0)\|$$

$$\le \sup_{0<\theta\le1} \|\partial_s\partial_s\mathcal{H}(s_0+\theta(s-s_0),y_0)\||s-s_0|^2$$

$$\quad + \sup_{0<\theta\le1} \|\partial_y\partial_y\mathcal{H}(s,y_0+\theta(y-y_0),y_0)\|\|y-y_0\|^2$$

$$\quad + \sup_{s\in[\tau,t]} \|\partial_s\partial_y\mathcal{H}(s,y_0)\||s-s_0|\|y-y_0|.$$

Proof of (3.7.23):

$$\sum_{k\notin\Gamma_\omega^n(A)\cup\Gamma_\omega^n(\Lambda)} C(\omega)\|Y_{\tau_{k+1}^n}(\omega) - Y_{\tau_k^n}(\omega)\|^2$$

$$\le 2C(\omega) \sum_{k\notin\Gamma_\omega^n(A)\cup\Gamma_\omega^n(\Lambda)} \|Z_{\tau_{k+1}^n}(\omega) - Z_{\tau_k^n}(\omega)\|^2$$

$$\leq 2C(\omega) \sup_{k \notin \Gamma^n_\omega(A) \cup \Gamma^n_\omega(\Lambda)} \left\| \int_{\tau^n_k}^{\tau^n_{k+1}} \int_A f(s, x, \omega) v(ds, dx) \right\|$$

$$\times \sum_{k=0}^{2^n-1} \left\| \int_{\tau^n_k}^{\tau^n_{k+1}} \int_A f(s, x, \omega) v(ds, dx) \right\|,$$

and since $f(s, x, \omega)$ is Bochner integrable $\mathbb{P}$-almost surely with respect to v and $v(ds, dx) = \alpha(ds)\beta(dx)$, with $\alpha(ds) \ll ds$, it follows that

$$\limsup_{n \to \infty} \sup_{k \notin \Gamma^n_\omega(A) \cup \Gamma^n_\omega(\Lambda)} \left\| \int_{\tau^n_k}^{\tau^n_{k+1}} \int_A f(s, x, \omega) v(ds, dx) \right\|$$

$$\leq \limsup_{n \to \infty} \sup_{k \notin \Gamma^n_\omega(A) \cup \Gamma^n_\omega(\Lambda)} \int_{\tau^n_k}^{\tau^n_{k+1}} \int_A \|f(s, x, \omega)\| v(ds, dx) = 0 \quad \mathbb{P}\text{-a.s.}$$

This completes the proof. Let us now assume that E is, in general, a Blackwell space. In this case the representation (3.7.17) for Z_t is no longer valid. However, also in this case, according to [49, Proposition II.1.14], there exist a sequence $(\int_j^A)_{j \in \mathbb{N}}$ of finite stopping times with $[\![\int_j^A]\!] \cap [\![\int_l^A]\!] = \emptyset$ for $j \neq l$ and an E-valued optional process ξ such that for every optional process $f : \Omega \times \mathbb{R}_+ \times E \to H$ with

$$\mathbb{P}\left(\int_0^t \int_A \|f(s, x)\| N(ds, dx) < \infty \right) = 1 \quad \text{for all } t \geq 0$$

we have the identity

$$\int_0^t \int_A f(s, x) N(ds, dx) = \sum_{k \in \mathbb{N}} f(\int_k^A, \xi_{\int_k^A}) \mathbb{1}_{\{\int_k^A \leq t\}}, \quad t \geq 0. \tag{3.7.25}$$

Let

$$\Gamma^n_\omega(A) := \{k \in 0, \dots, 2^n - 1 : \exists \int_j^A \in (\tau^n_k, \tau^n_{k+1}] :\}$$

with $\tau^n_k := \tau + \frac{k(t-\tau)}{2^n}$. Then equality (3.7.18) holds.

It can be shown, similarly to (3.7.19), that almost surely

$$
\lim_{n \to \infty} \sum_{k \in \Gamma_\omega^n(A) \cup \Gamma_\omega^n(\Lambda)} \mathcal{H}(\tau_{k+1}^n, Y_{\tau_{k+1}^n}(\omega)) - \mathcal{H}(\tau_k^n, Y_{\tau_k^n}(\omega))
$$

$$
= \lim_{n \to \infty} \sum_{k \in \Gamma_\omega^n(A)} \sum_j [\mathcal{H}(\tau_{k+1}^n, Y_{\tau_k^n}(\omega) + f(\int_j^A, \xi_{\int_j^A}))
$$

$$
- \mathcal{H}(\tau_k^n, Y_{\tau_k^n}(\omega))] \mathbf{1}_{\int_j^A \in (\tau_k^n, \tau_{k+1}^n]}
$$

$$
+ \lim_{n \to \infty} \sum_{k \in \Gamma_\omega^n(\Lambda)} \sum_j [\mathcal{H}(\tau_{k+1}^n, Y_{\tau_k^n}(\omega) + k(\int_j^\Lambda, \xi_{\int_j^\Lambda}))
$$

$$
- \mathcal{H}(\tau_k^n, Y_{\tau_k^n}(\omega))] \mathbf{1}_{\int_j^\Lambda \in (\tau_k^n, \tau_{k+1}^n]}
$$

$$
= \int_\tau^t \int_A \{\mathcal{H}(s, Y_{s-} + f(s, x)) - \mathcal{H}(s, Y_{s-})\} q(ds, dx)
$$

$$
+ \int_\tau^t \int_\Lambda \{\mathcal{H}(s, Y_{s-} + k(s, x)) - \mathcal{H}(s, Y_{s-})\} N(ds, dx)
$$

$$
+ \int_\tau^t \int_A \{\mathcal{H}(s, Y_{s-} + f(s, x)) - \mathcal{H}(s, Y_{s-})\} \nu(ds, dx).
$$

The proof of (3.7.20) is identical to the proof in the case where E is a separable Banach space. As in the previous case, it follows that (3.7.8) holds.   □

We now prove Theorem 3.7.2.

Proof Estimate (3.7.4) holds true by the continuity of the partial derivative $\partial_s H$, and (3.7.7) is valid because $\beta(\Lambda) < \infty$. By Taylor's theorem, the Cauchy–Schwarz. inequality and (3.7.1), we obtain $\mathbb{P}$-almost surely

$$
\int_0^t \int_A \|\mathcal{H}(s, Y_s + f(s, x)) - \mathcal{H}(s, Y_s)\|^2 \nu(ds, dx)
$$

$$
= \int_0^t \int_A \left\| \int_0^1 \partial_y \mathcal{H}(s, Y_s + \theta f(s, x)) f(s, x) d\theta \right\|^2 \nu(ds, dx)
$$

$$
\leq \int_0^t \int_A \int_0^1 \|\partial_y \mathcal{H}(s, Y_s + \theta f(s, x))\|^2 \|f(s, x)\|^2 d\theta \nu(ds, dx)
$$

$$
\leq \int_0^t \int_A \int_0^1 h_1(\|Y_s + \theta f(s, x)\|)^2 \|f(s, x)\|^2 d\theta \nu(ds, dx).
$$

Since h_1 is quasi-sublinear, for some constant $C > 0$ we get $\mathbb{P}$-almost surely

$$\int_0^t \int_A \|\mathcal{H}(s, Y_s + f(s, x)) - \mathcal{H}(s, Y_s)\|^2 \nu(ds, dx)$$

$$\leq C^2 \int_0^t \int_A \int_0^1 \left(h_1(\|Y_s\|) + Ch_1(\theta)h_1(\|f(s, x)\|)\right)^2 \|f(s, x)\|^2 d\theta\, \nu(ds, dx)$$

$$\leq 2C^2 \int_0^t \int_A h_1(\|Y_s\|)^2 \|f(s, x)\|^2 \nu(ds, dx)$$

$$+ 2C^4 h_1(1) \int_0^t \int_A h_1(\|f(s, x)\|)^2 \|f(s, x)\|^2 \nu(ds, dx) < \infty,$$

showing (3.7.5). By Taylor's theorem and (3.7.2), we obtain $\mathbb{P}$-almost surely

$$\int_0^t \int_A \|\mathcal{H}(s, Y_s + f(s, x)) - \mathcal{H}(s, Y_s) - \partial_y \mathcal{H}(s, Y_s) f(s, x)\| \nu(ds, dx)$$

$$= \int_0^t \int_A \left\| \int_0^1 \partial_{yy} \mathcal{H}(s, Y_s + \theta f(s, x))(f(s, x), f(s, x)) d\theta \right\| \nu(ds, dx)$$

$$\leq \int_0^t \int_A \int_0^1 \|\partial_{yy} \mathcal{H}(s, Y_s + \theta f(s, x))\| \, \|f(s, x)\|^2 d\theta\, \nu(ds, dx)$$

$$\leq \int_0^t \int_A \int_0^1 h_2(\|Y_s + \theta f(s, x)\|) \|f(s, x)\|^2 d\theta\, \nu(ds, dx).$$

Since h_2 is quasi-sublinear, for some constant $C > 0$ we get $\mathbb{P}$-almost surely

$$\int_0^t \int_A \|\mathcal{H}(s, Y_s + f(s, x)) - \mathcal{H}(s, Y_s) - \partial_y \mathcal{H}(s, Y_s) f(s, x)\| \nu(ds, dx)$$

$$\leq C \int_0^t \int_A \int_0^1 \left(h_2(\|Y_s\|) + Ch_2(\theta)h_2(\|f(s, x)\|)\right) \|f(s, x)\|^2 d\theta\, \nu(ds, dx)$$

$$\leq C \int_0^t \int_A h_2(\|Y_s\|) \|f(s,x)\|^2 \nu(ds, dx)$$

$$+ C^2 h_2(1) \int_0^t \int_A h_2(\|f(s,x)\|) \|f(s,x)\|^2 \nu(ds, dx) < \infty,$$

providing (3.7.6).

Let us prove that (3.7.8) holds. It is sufficient to prove (3.7.8) for the case where the Fréchét derivatives of $\mathcal{H}$ are uniformly bounded. In fact, if this is not the case, we can use the same method as in the proof of [73, Theorem 25.7]. We consider $\mathcal{H}_C \in C_b^{1,2}(\mathbb{R} \times F/G)$ such that, for $\|x\| \leq C, \mathcal{H}_C(s, x)$ coincides, with all its derivatives, with $\mathcal{H}(s, x)$. We prove the Itô formula for the process $(Y_{t \wedge \tau_C})_{t \in [0,T]}$, where τ_C is the stopping time with

$$\tau_C(\omega) = \inf_{t \in [0,T]} \{\|Y_t\| > C\}.$$

Note that the probability that Y_t has a jump at time t is zero, as Y_t is continuous in probability, so that $\mathcal{H}_C(s, Y_{s \wedge \tau_C}) = \mathcal{H}(s, Y_{s \wedge \tau_C})$, and similarly for the derivatives. It follows that $(y_{t \wedge \tau_C})_{t \in [0,T]}$ satisfies the Itô formula (3.7.8). As a consequence of properties (3.7.4)–(3.7.7), taking the limit as $C \to \infty$, all terms converge to the terms in (3.7.8), where Theorem 3.5.17 is used to prove that the stochastic integral term also converges. It follows that $(y_t)_{t \in [0,T]}$ solves (3.7.8).

From properties (3.7.4)–(3.7.7), it follows that it is sufficient to prove the theorem for the case where $f(t, x, \omega)$ is a simple function. By Lemma 3.7.5, (3.7.8) is proven.   □

Example 3.7.6 Suppose that F is a separable Hilbert space. Then $H(x) = \|x\|^2$ is of class $C^2(F; \mathbb{R})$ with

$$H_x(x)v = 2\langle x, v \rangle \quad \text{and} \quad H_{xx}(x)(v, w) = 2\langle v, w \rangle.$$

Therefore, we have

$$\|H_x(x)\| \leq 2\|x\| \quad \text{and} \quad \|H_{xx}(x)\| \leq 2.$$

Consequently, if

$$\int_0^t \int_B \|f(s,x)\|^2 \nu(ds, dx) + \int_0^t \int_B \|f(s,x)\|^4 \nu(ds, dx) < \infty \quad \text{for all } t \in \mathbb{R}_+,$$

then Theorem 3.7.2 applies and yields that $\mathbb{P}$-almost surely

$$\|Y_t\|^2 = \|Y_0\|^2 + \int\limits_0^t \int\limits_B \left(\|Y_{s-} + f(s,x)\|^2 - \|Y_{s-}\|^2 \right) q(ds,dx)$$

$$+ \int\limits_0^t \int\limits_B \left(\|Y_s + f(s,x)\|^2 - \|Y_s\|^2 - 2\langle Y_s, f(s,x)\rangle \right) \nu(ds,dx)$$

$$+ \int\limits_0^t \int\limits_{B^c} \left(\|Y_{s-} + g(s,x)\|^2 - \|Y_{s-}\|^2 \right) N(ds,dx), \quad t \geq 0.$$

3.8 Remarks and Related Literature

The Wiener integral derivation from condition (3.1.4) was given in [91], along with Theorem 3.1.13. The idea of this originated in [90]. (See also [97] for the real-valued case.) The concepts connected with Lévy processes are taken from [94] and the derivation of the Lévy–Itô decomposition in Banach spaces combines ideas of [94] with the work of [3], where the infinite-dimensional version is proved under the type 2 condition. Earlier work of [21] defined the integral with respect to compensated Poisson random measures differently, following the lines of [45] instead of [97], to obtain the Lévy–Itô decomposition. That both definitions of integrals are equivalent is discussed in [91].

The definition of the Itô integral for M-type 2 spaces appears in [39, 65] (and previously, under restricted conditions, in [91]). We take the generalization under (3.5.7) given here from [67], and the Pettis integral in M-type 2 spaces follows the approach in [67] for integrating non-anticipating processes. Integration with respect to martingales taking values in M-type 2 spaces in given by [86]. As our integral is a martingale in an M-type 2 Banach space, we connect integrals with respect to these martingales to integrals with respect to compensated Poisson random measures. This material originally appeared in [66].

The results in Sect. 3.4 concerning isomorphism of L^p-spaces originated in [92] and for particular cases appears in the classical literature, e.g., [20, 58].

Itô's formula was originally given in this context in [93]. However, there it was only proven for a smaller class of functions. Here, we have given an improvement of [93], which will also appear in [68].

In [39], the Itô stochastic integral was given in M-type p Banach spaces instead of M-type 2 Banach spaces. Our technique can be used to obtain similar results. However our condition (3.5.9) allows us to define the Itô integral in any Banach space without involving geometry. One involves geometry in connecting the Pettis integral to the Itô integral using Rosinski's method. This method was exploited by J. M. A. M. Van Neerven and his collaborators in the Gaussian case using an idea of D. J. H. Garling in a series of papers. One can find them on MathSciNet. Because they are too numerous and our work involves condition (3.5.9) and jump processes, we do not refer to these results here.

Chapter 4
Stochastic Integral Equations in Banach Spaces

In this chapter, we first study the solutions of stochastic differential equations with non-Markovian Lipschitz condition and growth condition. In this case the drift and noise coefficients $a(t, Z)$ and $f(t, x, Z)$, being non-anticipating, depend on the path of the solution $Z = (Z_t)_{t \in [0,T]}$. This is done by defining the equation on a probability space with $\Omega := D(\mathbb{R}_+, E)$, the space of càdlàg functions on $\mathbb{R}_+ \to E$, and with the σ-algebra generated by the cylinder sets of $D(\mathbb{R}_+, E)$, where E is a separable Banach space.

After proving the existence and uniqueness in this case, we consider on a general probability space a stochastic differential equation with coefficients defined on $\mathbb{R}_+ \times F \times \Omega$ and $\mathbb{R}_+ \times F \times E \times \Omega$ with values in F. The coefficients in this case depend on the value of the solution Z at time t, i.e. the coefficients appearing in the drift and noise are of the form $a(t, Z_t, \omega)$ and $f(t, x, Z_t, \omega)$.

4.1 Existence and Uniqueness Results for Non-Markovian Solutions

Let us denote by $\mathcal{C}_t$ the σ-field on $D(\mathbb{R}_+, E)$ generated by cylinder sets with basis in $[0, t]$. We consider the stochastic differential equation

$$dZ_t = a(t, Z)dt + \int_E f(t, x, Z)q(dt, dx) \qquad (4.1.1)$$

$$Z_0 = \Phi$$

with $\Phi : \mathbb{R}_+ \to E$.

We assume that a and f are non-linear functions with

$$a : \mathbb{R}_+ \times D(\mathbb{R}_+, E) \to E$$

$$f : \mathbb{R}_+ \times E \times D(\mathbb{R}_+, E) \to E$$

© Springer International Publishing Switzerland 2015
V. Mandrekar and B. Rüdiger, *Stochastic Integration in Banach Spaces*,
Probability Theory and Stochastic Modelling 73, DOI 10.1007/978-3-319-12853-5_4

where

(A1) $f(t, x, z)$ is jointly measurable and, for each $t \in \mathbb{R}_+$, $x \in E$, $f(t, x, \cdot)$ is $\mathcal{C}_t$ adapted.
(A2) $a(t, x)$ is jointly measurable and, for each $t \in \mathbb{R}_+$, $a(t, \cdot)$ is $\mathcal{C}_t$ adapted.

For each $t \in \mathbb{R}_+$ we consider the function

$$\theta_t : D(\mathbb{R}_+, E) \rightarrow D(\mathbb{R}_+, E)$$
$$z \rightarrow \theta_t(z)$$

with

$$\theta_t(z)(s) = z_s \quad for\ 0 \leq s < t$$
$$= z_t \quad for\ s \geq t$$

and assume $f(t, x, z) = f(t, x, \theta_t(z))$ and $a(t, z) = a(t, \theta_t(z))$.
   We further assume

(A3) Let $T > 0$ be fixed. Then there exists an $l > 0$ such that for $t_1, t_2 \in [0, T]$

$$\int_{t_1}^{t_2} \int_E \|f(t, x, z)\|_E^2 \beta(dx) dt + \int_{t_1}^{t_2} \|a(t, z)\|_E^2 dt \leq l \int_{t_1}^{t_2} (1 + \|\theta_t(z)\|_\infty^2) dt$$

where $\|z\|_\infty = \sup_{0 \leq t \leq T} \|z(s)\|_E$.
   Moreover, we assume that there is a constant K_β such that inequality (3.5.7) is satisfied, so that, due to Lemma 3.5.5, the Itô integral in (4.1.1) is well defined, and inequality (3.5.9) holds.
   Now we define

$$I(t, Z) := \int_0^t a(s, Z) ds + \int_0^t \int_E f(s, x, Z) q(ds, dx).$$

   Then we get

Lemma 4.1.1 *There exists a constant $C_{l,T}$ such that for any $\mathcal{C}_t$-stopping time τ and $t \in [0, T]$*

$$\mathbb{E}\left[\sup_{0 \leq s \leq t \wedge \tau} \|I(s, Z)\|_E^2 \right] \leq C_{l,T}\left(t + \int_0^t \mathbb{E}[\sup_{0 \leq v \leq s \wedge t} \|Z_v\|_E^2] ds \right).$$

Proof Note that

$$\sup_{0 \le s \le t \wedge \tau} \|I(s, Z)\|_E^2 \le 2 \sup_{0 \le s \le t \wedge \tau} \left\| \int_0^s a(v, Z) dv \right\|_E^2$$

$$+ 2 \sup_{0 \le s \le t \wedge \tau} \left\| \int_0^s \int_E f(v, u, Z) q(dv, du) \right\|_E^2$$

$$\mathbb{E} \left[\sup_{0 \le s \le t \wedge \tau} \left\| \int_0^s a(v, Z) dv \right\|_E^2 \right] \le t \mathbb{E} \left[\sup_{0 \le s \le t} \left(l \int_0^s (1 + \|\theta_v(Z)\|_\infty^2) dv \right)^2 \right].$$

(4.1.2)

Using the martingale property of the second term and Doob's inequality we get

$$\mathbb{E} \left[\sup_{0 \le s \le t \wedge \tau} \left\| \int_0^s \int_E f(v, u, Z) q(dv, du) \right\|_E^2 \right] \le C_{l,T} \left(t + \mathbb{E} \left[\int_0^t \|\theta_v(Z)\|_\infty^2 dv \right] \right).$$

(4.1.3)

In fact,

$$\mathbb{E} \left[\sup_{0 \le s \le t \wedge \tau} \left\| \int_0^s \int_E f(v, u, Z) q(dv, du) \right\|_E^2 \right] \le \mathbb{E} \left[\left\| \int_0^{t \wedge \tau} \int_E f(v, u, Z) q(dv, du) \right\|_E^2 \right]$$

$$\le K_\beta \mathbb{E} \left[\int_0^t \|f(v, u, Z)\|_E^2 \beta(du) dv \right]$$

$$\le K_\beta l \mathbb{E} \left[\int_0^t (1 + \|\theta_v(Z)\|_\infty^2) dv \right]$$

$$\le K_\beta l \left(t + \mathbb{E} \left[\int_0^t \|\theta_v(Z)\|_\infty^2 dv \right] \right).$$

The result follows from inequalities (4.1.2) and (4.1.3). $\qquad \square$

Let $T > 0$ and define

$$\mathcal{H}_2^T = \xi := (\xi_s)_{s \in [0,T]} : \xi_s(\omega) \text{ is jointly measurable, } \mathcal{C}_t\text{-adapted, with } \mathbb{E}[\sup_{0 \le s \le t} \|\xi_s\|^2]$$

$< \infty\}.$

We have just proved that

$$I : \mathcal{H}_2^T \to \mathcal{H}_2^T$$
$$\xi \to I(\cdot, \xi).$$

Lemma 4.1.2 *The linear space* $\mathcal{H}_2^T$, *equipped with the norm*

$$\|Z\|_{\mathcal{H}_2^T} = \mathbb{E}\left[\sup_{t \in [0,T]} \|Z_t\|^2 \right]^{1/2},$$

is a Banach space.

The proof is given for the more general statement Lemma 4.2.1 in the next section. Let us now assume the Lipschitz condition

(A4) Let $T > 0$ be fixed. Then there exists a $K > 0$ such that for fixed $t_1, t_2 \in \mathbb{R}$ and $Z, Y \in D(\mathbb{R}_+, E)$

$$\int_{t_1}^{t_2} \int_E \|f(t, x, Z) - f(t, x, Y)\|_E^2 \beta(dx) dt + \int_{t_1}^{t_2} \|a(t, Z) - a(t, Y)\|_E^2 dt$$

$$\leq l \int_{t_1}^{t_2} \|\theta_t(Z) - \theta_t(Y)\|_\infty^2 dt.$$

Lemma 4.1.3 *The map* $I : \mathcal{H}_2^T \to \mathcal{H}_2^T$ *is continuous. There exists a* $C_{K,T}$ *depending on* K, T *such that*

$$\mathbb{E}[\sup_{0 \leq s \leq T} \|I(s, Z^1) - I(s, Z^2)\|_E^2] \leq C_{K,T} \int_0^T \mathbb{E}[\sup_{0 \leq s \leq T} \|Z_s^1 - Z_s^2\|_E^2 ds].$$

Exercise: Use condition (A4) and follow the proof of Lemma 4.1.1 to get Lemma 4.1.3.

Theorem 4.1.4 *Let* $T > 0$, $z \in E$. *There exists a unique solution* $Z = (Z_s)_{s \in [0,T]}$ *in* $\mathcal{H}_2^T$ *which satisfies*

$$Z_t = z + \int_0^t a(s, Z) ds + \int_0^t \int_E f(s, x, Z) q(ds, dx).$$

Proof We shall prove that the solution can be approximated in $\mathcal{H}_2^T$ by $Z^n = (Z_s^n)_{s \in [0,T]}$ when $n \to \infty$. Define for $n \in \mathbb{N}$

$$Z_s^0(\omega) = z \quad P - a.s.$$
$$Z_s^{n+1}(\omega) = I(s, Z^n(\omega)).$$

Observe that $(Z_t^n)_t \in [0, T]$ is $\mathcal{F}_t$-adapted. Define

$$v_t^n = \mathbb{E}[\sup_{0 \leq s \leq t} \|Z_s^{n+1} - Z_s^n\|_E^2].$$

By Lemma 4.1.1 we get

$$v_t^0 = \mathbb{E}\left[\int_0^t \sup_{0\le s\le t} \|Z_s^1 - Z_s^0\|_E^2 ds\right] \le V_{l,T}(z)$$

with $V_{l,T}(z) = C_{l,T} 2T\|z\|^2$. Also by Lemma 4.1.3

$$v_t^1 \le \int_0^t \mathbb{E}[\sup_{0\le s\le t} \|Z_s^1 - Z_s^0\|_E^2] ds \le \frac{T^2 C_{K,T}^2}{2} V_{l,T}(z). \tag{4.1.4}$$

By induction

$$v_t^n \le C_{K,T} \int_0^t v_s^{n-1} ds \le \frac{T^{n+1} C_{K,T}^{n+1}}{(n+1)!} V_{l,T}(z). \tag{4.1.5}$$

Let $\epsilon_n = \left(\frac{T^{n+1} C_{K,T}^{n+1}}{(n+1)!}\right)^{\frac{1}{3}}$. Then by Chebychev's inequality

$$\mathbb{P}(\sup_{0\le s\le t} \|Z_t^{n+1} - Z_t^n\|_E^2 > \epsilon_n) \le \epsilon_n^2 V_{l,T}(z).$$

As $\sum_n \epsilon_n^2$ is convergent, we get by the Borel–Cantelli Lemma that $\sum_{n=1}^{\infty} \sup_{0\le s\le t} \|Z_t^{n+1} - Z_t^n\|_E^2$ converges $\mathbb{P}$-a.s.

This gives that Z^n converges to some process $Z = (Z_t)_{t\in[0,T]}$ in the supremum norm, where $Z \in D([0,T], E)$.

Moreover

$$\mathbb{E}[\sup_{0\le t\le T} \|Z_t - Z_t^n\|^2] = \mathbb{E}[\lim_{n\to\infty} \sup_{0\le t\le T} \|\sum_{k=n}^{n+m-1} (Z_t^{k+1} - Z_t^k)\|_E^2]$$

$$\le \mathbb{E}\left[\lim_{n\to\infty}\left(\sum_{k=n}^{n+m-1} \|Z_t^{k+1} - Z_t^k\|_E^2 k\frac{1}{k}\right)^2\right]$$

and by Schwarz's inequality

$$\mathbb{E}[\sup_{0\le t\le T} \|Z_t - Z_t^n\|^2] \le \sum_{k=n}^{\infty} \mathbb{E}[\sup_{0\le t\le T} \|Z_t^{k+1} - Z_t^k\|_E^2 k^2] \sum_{k=n}^{\infty} \frac{1}{k^2}$$

$$\le V_{l,T}(z)\left(\sum_{k=n}^{\infty} \frac{T^{k+1} C_{K,T}^{k+1}}{(k+1)!} k^2\right)\left(\sum_{k=n}^{\infty} \frac{1}{k^2}\right)$$

which converges to zero as $n \to \infty$. That is, $Z^n \to Z$ in $\mathcal{H}_2^T$.

By Lemma 4.1.3 and

$$Z_s^{n+1} = I(s, Z^n(\omega))$$

with $Z^n \to Z$ a.e., we get that Z, obtained by contraction in $\mathcal{H}_2^T$, satisfies (4.1.1).

To prove uniqueness, suppose $(Z_t)_{t\in[0,T]}$ and $(Y_t)_{t\in[0,T]}$ are solutions to (4.1.1). Let

$$v_t = \mathbb{E}[\sup_{0 \le s \le t} \|Z_s - Y_s\|_E^2].$$

Then, as in (4.1.4), we have

$$v_t \le C_{K,T} \int_0^t v_s ds,$$

and by induction $v_t \le \frac{C_{K,T}^n}{n!} \mathbb{E}[\sup_{0 \le s \le t} \|Z_s - Y_s\|_E^2]$ which tends to zero as $n \to \infty$.
Thus we have $v_t = 0 \; \forall t \in [0, T]$, and hence uniqueness.   □

4.2 Existence and Uniqueness of Markovian Solutions in $D([0, T], F)$

We now consider equations defined on any filtered probability space $(\Omega, \mathcal{F}, \mathcal{F}_t, \mathbb{P})$, satisfying the usual conditions, with values on a separable Banach space F. As usual the mark space $(E, \mathcal{E})$ defining the cPrm is a Blackwell space. We consider the stochastic differential equation (SDE)

$$Z_t = \Phi_t + \int_0^t a(s, Z_s)ds + \int_0^t \int_E f(s, x, Z_s)q(ds, dx) \qquad (4.2.1)$$

under conditions

(B1) $f(t, x, z, \cdot)$ is a $\mathcal{B}([0, T] \times E \times F) \otimes \mathcal{F}_t/\mathcal{B}(F)$-measurable function such that for fixed $t \in [0, T]$, $x \in E$ and $z \in F$, $f(t, x, z, \cdot)$ is $\mathcal{F}_t$ adapted.

(B2) $a(t, x, \cdot)$ is a $\mathcal{B}([0, T]) \times E \otimes \mathcal{F}_t/\mathcal{B}(F)$-measurable function such that for fixed $t \in [0, T]$ and $x \in E$, $a(t, x, \cdot)$ is $\mathcal{F}_t$ adapted.

(B3) There exists a constant $L > 0$ such that

$$T\|a(t, z) - a(t, y)\|_F^2 + \int_E \|f(t, x, z) - f(t, x, y)\|_F^2 \beta(dx)$$

$$\le L\|z - y\|^2 \quad \text{for all } t \in [0, T], \; z, y \in F \quad \mathbb{P} - a.e.$$

(B4) There exists a constant $M > 0$ such that

$$\|a(t, z)\|_F^2 + \int_E \|f(t, x, z)\|_F^2 \beta(dx)$$
$$\leq M(\|z\|^2 + 1) \quad \text{for all } t \in [0, T], z \in F \quad \mathbb{P} - a.e.$$

Moreover, we assume that there is a constant K_β such that inequality (3.5.7) is satisfied, so that, due to Lemma 3.5.5, the Itô integral in (4.2.1) is well defined, and inequality (3.5.9) holds.

Let S_T^2 be the linear space of all càdlàg adapted processes Z such that

$$\mathbb{E}\left[\sup_{t \in [0,T]} \|Z_t\|_F^2 \right] < \infty,$$

where we identify processes whose paths coincide almost surely. Note that, by the completeness of the filtration, adaptedness does not depend on the choice of the representative.

Lemma 4.2.1 *The linear space S_T^2, equipped with the norm*

$$\|Z\|_{S_T^2} = \mathbb{E}\left[\sup_{t \in [0,T]} \|Z_t\|^2 \right]^{1/2},$$

is a Banach space.

Proof Let $(Z^n)_{n \in \mathbb{N}}$ be a Cauchy sequence in S_T^2. Then there exists a subsequence $(n_k)_{k \in \mathbb{N}}$ such that

$$\mathbb{E}\left[\sup_{t \in [0,T]} \|Z_t^{n_k} - Z_t^{n_{k+1}}\|^2 \right] < \frac{2^{-k}}{k^2}, \quad k \in \mathbb{N}.$$

By the monotone convergence theorem, we deduce that

$$\mathbb{E}\left[\sum_{k=1}^{\infty} k^2 \sup_{t \in [0,T]} \|Z_t^{n_k} - Z_t^{n_{k+1}}\|^2 \right] = \sum_{k=1}^{\infty} k^2 \mathbb{E}\left[\sup_{t \in [0,T]} \|Z_t^{n_k} - Z_t^{n_{k+1}}\|^2 \right] < 1, \quad k \in \mathbb{N}.$$
(4.2.2)

Therefore, there exists an $\Omega_0 \in \mathcal{F}$ with $\mathbb{P}(\Omega_0) = 1$ such that

$$\sum_{k=1}^{\infty} k^2 \sup_{t \in [0,T]} \|Z_t^{n_k}(\omega) - Z_t^{n_{k+1}}(\omega)\|^2 < \infty, \quad \omega \in \Omega_0.$$

Fix $\omega \in \Omega_0$ and let $\epsilon > 0$ be arbitrary. There exists an index $k_0 \in \mathbb{N}$ such that

$$\sum_{k=l}^{m-1} k^2 \sup_{t \in [0,T]} \|Z_t^{n_k}(\omega) - Z_t^{n_{k+1}}(\omega)\|^2 < \frac{6\epsilon^2}{\pi^2} \quad \text{for all } l, m \geq k_0 \text{ with } l \leq m.$$

Therefore, for all $l, m \geq k_0$ with $l \leq m$ we obtain, by using the Cauchy–Schwarz inequality,

$$\sup_{t \in [0,T]} \|Z_t^{n_l}(\omega) - Z_t^{n_m}(\omega)\| \leq \sum_{k=l}^{m-1} \sup_{t \in [0,T]} \|Z_t^{n_k}(\omega) - Z_t^{n_{k+1}}(\omega)\|$$

$$\leq \left(\sum_{k=l}^{m-1} \frac{1}{k^2} \right)^{1/2} \left(\sum_{k=l}^{m-1} k^2 \sup_{t \in [0,T]} \|Z_t^{n_k}(\omega) - Z_t^{n_{k+1}}(\omega)\|^2 \right)^{1/2} < \epsilon.$$

Consequently, for almost all $\omega \in \Omega$ the sequence $(Z^{n_k}(\omega))_{k \in \mathbb{N}}$ is a Cauchy sequence in the Banach space of all càdlàg functions from $[0, T]$ to F, equipped with the supremum norm. Thus, there exists a càdlàg process Z such that almost surely

$$\sup_{t \in [0,T]} \|Z_t^{n_k} - Z_t\| \to 0,$$

whence the process Z is also adapted. For each $k \in \mathbb{N}$ we have almost surely

$$\|Z_t^{n_k} - Z_t\|^2 = \lim_{l \to \infty} \|Z_t^{n_k} - Z_t^{n_l}\|^2 \leq \left(\sum_{l=1}^{\infty} \|Z_t^{n_l} - Z_t^{n_{l+1}}\| \right)^2$$

$$\leq \left(\sum_{l=1}^{\infty} \frac{1}{l^2} \right) \left(\sum_{l=1}^{\infty} l^2 \|Z_t^{n_l} - Z_t^{n_{l+1}}\|^2 \right), \quad t \in [0, T]$$

and hence

$$\sup_{t \in [0,T]} \|Z_t^{n_k} - Z_t\|^2 \leq \frac{\pi^2}{6} \sum_{l=1}^{\infty} l^2 \sup_{t \in [0,T]} \|Z_t^{n_l} - Z_t^{n_{l+1}}\|^2.$$

By (4.2.2), we obtain

$$\|Z\|_{S_T^2} \leq \|Z^1\|_{S_T^2} + \|Z^1 - Z\|_{S_T^2} < \infty,$$

showing $Z \in S_T^2$, and, by using Lebesgue's theorem, we have $Z^{n_k} \to Z$ in S_T^2.

Let $\epsilon > 0$ be arbitrary. There exists an index $k_0 \in \mathbb{N}$ such that $\|Z^{n_k} - Z\|_{S_T^2} < \frac{\epsilon}{2}$ for all $k \geq k_0$ and $\|Z^n - Z^m\|_{S_T^2} < \frac{\epsilon}{2}$ for all $n, m \geq k_0$. Therefore

$$\|Z^n - Z\|_{S_T^2} \leq \|Z^n - Z^{n_k}\|_{S_T^2} + \|Z^{n_k} - Z\|_{S_T^2} < \epsilon.$$

Consequently, we have $Z^n \to Z$ in S_T^2. $\square$

Theorem 4.2.2 *Suppose Assumptions (B1), (B2), (B3) and (B4) are fulfilled,* $\Phi|_{[0,T]} \in S_T^2$ *for all* $T \geq 0$ *and that the Banach space F is of M-type 2. Then, there exists a unique solution Z for the integral equation (4.2.1) such that $Z|_{[0,T]} \in S_T^2$ for all* $T \geq 0$.

Proof Due to Assumption (B1), it suffices to prove existence and uniqueness on S_T^2 for each $T \geq 0$. In the sequel let $T \geq 0$ be arbitrary. We divide the proof into the following steps:

1. For any $Z \in S_T^2$ we define the process SZ by

$$(SZ)_t := \Phi_t + \int_0^t a(s, Z_s)ds + \int_0^t \int_E f(s, x, Z_s)q(ds, dx), t \in [0, T].$$

We shall first prove that the process SZ is well defined. Indeed, by the linear growth condition (B4) we have

$$\mathbb{E}\left[\int_0^T \|a(s, Z_s)\|^2 ds\right] \leq M\mathbb{E}\left[\int_0^T (\|Z_s\|^2 + 1)ds\right] = M\left(T + \mathbb{E}\left[\int_0^T \|Z_s\|^2 ds\right]\right)$$

$$\leq MT\left(1 + \mathbb{E}\left[\sup_{s\in[0,T]} \|Z_s\|^2\right]\right)$$

$$< \infty,$$

showing that $a(s, Z) \in L_T^2(F)$, as well as

$$\mathbb{E}\left[\int_0^T \int_E \|f(s, x, Z_s)\|^2 \beta(dx)ds\right] \leq M\mathbb{E}\left[\int_0^T (\|Z_s\|^2 + 1)ds\right]$$

$$\leq M\left(T + \mathbb{E}\left[\int_0^T \|Z_s\|^2 ds\right]\right)$$

$$\leq 2M^2 T\left(1 + \mathbb{E}\left[\sup_{s\in[0,T]} \|Z_s\|^2 ds\right]\right) < \infty,$$

showing that $f(s, x, Z) \in L_{T,\beta}^2(E, F)$.

2. Next, we show that $SZ \in S_T^2$ for all $Z \in S_T^2$. By Doob's inequality (Theorem 2.3.5), estimate (3.5.17) and the linear growth condition (B4) we obtain

$$\mathbb{E}\left[\sup_{t\in[0,T]} \|(SZ)_t\|^2\right] \leq 3\mathbb{E}\left[\sup_{t\in[0,T]} \|\Phi_t\|^2\right] + 3\mathbb{E}\left[\sup_{t\in[0,T]} \left\|\int_0^t a(s, Z_s)ds\right\|^2\right]$$

$$+ 3\mathbb{E}\left[\sup_{t\in[0,T]} \left\|\int_0^T \int_E f(s, x, Z_s)q(ds, dx)\right\|^2\right]$$

$$\leq 3\mathbb{E}\left[\sup_{t\in[0,T]} \|\Phi_t\|^2\right] + 3\mathbb{E}\left[\left(\int_0^T \|a(s, Z_s)\|ds\right)^2\right]$$

$$+ 12\mathbb{E}\left[\left\|\int_0^T \int_E f(s, x, Z_s)q(ds, dx)\right\|^2\right]$$

$$\leq 3\mathbb{E}\left[\sup_{t\in[0,T]} \|\Phi_t\|^2\right] + 3T\mathbb{E}\left[\int_0^T \|a(s, Z_s)\|^2 ds\right]$$

$$+ 12K^2\mathbb{E}\left[\int_0^T \int_E \|f(s, x, Z_s)\|^2 \beta(dx)ds\right]$$

$$\leq 3\mathbb{E}\left[\sup_{t\in[0,T]} \|\Phi_t\|^2\right] + 3(T + 4K^2)M\mathbb{E}\left[\int_0^T (\|Z_s\|^2 + 1)dsdt\right]$$

$$\leq 3\mathbb{E}\left[\sup_{t\in[0,T]} \|\Phi_t\|^2\right] + 3(T + 4K^2)MT\left(T + \mathbb{E}\left[\int_0^T \|Z_s\|^2 ds\right]\right)$$

$$\leq 3\mathbb{E}\left[\sup_{t\in[0,T]} \|\Phi_t\|^2\right] + 3(T + 4K^2)MT^2\left(1 + \mathbb{E}\left[\sup_{s\in[0,T]} \|Z_t\|^2\right]\right) < \infty.$$

Therefore, the operator $\mathcal{S}$ maps S_T^2 into itself.

3. Now, we shall prove that $\mathcal{S}$ has a unique fixed point. For two arbitrary processes $Y, Z \in S_T^2$, Doob's inequality (Theorem 2.3.5), estimate (3.5.17) and the Lipschitz condition (B3) gives us for all $t \in [0, T]$ the estimate

$$\mathbb{E}\left[\sup_{s\in[0,t]} \|(\mathcal{S}Y)_s - (\mathcal{S}Z)_s\|^2\right] \leq 2\mathbb{E}\left[\sup_{s\in[0,t]} \left\|\int_0^s (a(v, Y_v) - a(v, Z_v))dv\right\|^2\right]$$

$$+ 2\mathbb{E}\left[\sup_{s\in[0,t]} \left\|\int_0^s \int_E (f(v, x, Y_v)\right.\right.$$

$$\left.\left. - f(v, x, Z_v))q(dv, dx)\right\|^2\right]$$

$$\leq 2\mathbb{E}\left[\left(\int_0^t \|a(s, Y_s) - a(s, Z_s)\|ds\right)^2\right]$$

$$+ 8\mathbb{E}\left[\left\|\int_0^t \int_E (f(s, x, Y_s) - f(s, x, Z_s))q(ds, dx)\right\|^2\right]$$

$$\leq 2T\mathbb{E}\left[\int_0^t \|a(s, Y_s) - a(s, Z_s)\|^2 ds\right]$$

$$+ 8K^2\mathbb{E}\left[\int_0^t \int_E \|f(s, x, Y_s) - f(s, x, Z_s)\|^2 \beta(dx)ds\right]$$

$$\leq 2(T + 4K^2)L\mathbb{E}\left[\int_0^t \|Y_s - Z_s\|^2 ds\right]$$

$$= 2(T + 4K^2)L\int_0^t \mathbb{E}[\|Y_s - Z_s\|^2]ds$$

$$\leq 2(T + 4K^2)L\int_0^t \mathbb{E}\left[\sup_{v\in[0,s]} \|Y_v - Z_v\|^2\right]ds.$$

By induction, for each $n \in \mathbb{N}$ we deduce that

$$
\mathbb{E}\left[\sup_{t \in [0,T]} \|\mathcal{S}^n Y_t - \mathcal{S}^n Z_t\|^2\right]
$$

$$
\leq (2(T + 4K^2)L)^n \int_0^T \int_0^{t_1} \cdots \int_0^{t_{n-2}} \left(\int_0^{t_{n-1}} \mathbb{E}\left[\sup_{v \in [0,s]} \|Y_s - Z_s\|^2\right] ds\right) dt_{n-1} \ldots dt_2 dt_1
$$

$$
\leq (2(T + 4K^2)L)^n \left(\int_0^T \int_0^{t_1} \cdots \int_0^{t_{n-1}} 1 dt_n \ldots dt_2 dt_1\right) \mathbb{E}\left[\sup_{t \in [0,T]} \|Y_t - Z_t\|^2\right]
$$

$$
\leq (2(T + 4K^2)L)^n \frac{T^n}{n!} \mathbb{E}\left[\sup_{t \in [0,T]} \|Y_t - Z_t\|^2\right].
$$

Hence, there exists an $n \in \mathbb{N}$ such that $\mathcal{S}^n : S_T^2 \to S_T^2$ is a contraction. Taking into account Lemma 4.2.1, this implies that the mapping $\mathcal{S}$ has a unique fixed point.

$\square$

4.3 Existence and Uniqueness of Markovian Solutions Under Local Lipschitz Conditions

This section deals again with existence and uniqueness of the solution of (4.2.1) assuming conditions (B1), (B2) and (B4). Instead of the Lipschitz condition (B3) we assume the more general local Lipschitz condition (B3′) below

(B3′) for each $n \in \mathbb{N}$ there exists a constant $L(n) > 0$ such that

$$
T\|a(t, z) - a(t, y)\|_F^2 + \int_E \|f(t, x, z) - f(t, x, y)\|_F^2 \beta(dx)
$$

$$
\leq L(n)\|z - y\|^2 \quad \text{for all } t \in [0, T], \ z, y \in F, \ \|z\| < n, \ \|y\| < n \quad P - a.e.
$$

Theorem 4.3.1 *Suppose Assumptions (B1), (B2), (B3′) and (B4) are fulfilled, $\Phi|_{[0,T]} \in S_T^2$ for all $T \geq 0$ and that the Banach space F is of M-type 2. Then there exists a solution Z of the integral equation (4.2.1) such that $Z|_{[0,T]} \in S_T^2$ for all $T \geq 0$. The solution is $P \otimes \lambda$-a.e. unique on $\Omega \times \mathbb{R}_+$.*

Proof
Step 1. For $n \in \mathbb{N}$ let

$$
a_n(s, z) = a\left(s, \frac{z}{1 + d(z, B_n)}\right) \tag{4.3.1}
$$

$$
f_n(s, x, z) = f\left(s, x, \frac{z}{1 + d(z, B_n)}\right), \tag{4.3.2}
$$

where we denote with $B_n = B(0, n)$ the centered ball in F with radius n, and with $d(z, B_n)$ the distance of $z \in F$ from B_n, then a_n, f_n satisfy (B1), (B2), (B3), (B4). Hence for each $n \in \mathbb{N}$ there exists a unique solution Z^n of the integral equation

$$Z_t^n = \Phi_t + \int_0^t a_n(s, Z_s^n)ds + \int_0^t \int_E f_n(s, x, Z_s^n)q(ds, dx). \qquad (4.3.3)$$

Step 2.   For $n \in \mathbb{N}$ we define the stopping time

$$\tau_n = \inf\{t \in [0, T] : \|Z_t^n\| > n\}.$$

By uniqueness of solutions for (4.3.1) we get a.s.

$$Z_t^n = Z_{t+1}^n \quad \text{for} \quad t \in [0, \tau_n)$$

giving $\mathbb{P}(\tau_n \leq \tau_{n+1}) = 1$.
   Our goal is to prove

$$\mathbb{P}(\cup_{n \in \mathbb{N}}\{\tau_n = T\}) = 1. \qquad (4.3.4)$$

Then $Z = \lim_n Z^n$ is the desired solution of (4.3.1). Let n be arbitrary. By inequality (3.5.17) and (B4) we get

$$\mathbb{E}[\|Z_t^n\|^2] \leq 16\mathbb{E}[\|\Phi_t\|^2 + 16\mathbb{E}\left[\int_0^t \|a_n(s, Z_s^n)\|^2 ds\right]$$

$$\times 16K^2\mathbb{E}\left[\int_0^t \int_E \|f_n(s, x, Z_s^n)\|^2 \beta(dx)ds\right]$$

$$\leq 16\mathbb{E}[\|\Phi_t\|^2 + 16M(1 + K^2)\left(T + \int_0^t \|Z_s^n\|^2 ds\right).$$

By Gronwall's Lemma for $t \in [0, T]$

$$\mathbb{E}[\|Z_t^n\|^2] \leq \left(16\mathbb{E}[\|\Phi_t\|^2] + 16M(1 + K^2)T\right)e^{16M(1+K^2)T}. \qquad (4.3.5)$$

Therefore by Markov's inequality and Doob's inequality (Theorem 2.3.5), estimate (3.5.17) and the linear growth condition (B4).

$$\mathbb{P}(\tau_n < T) \leq \mathbb{P}(\sup_{t \in [0,T]} \|Z_t^n\| > n)$$

$$\leq \mathbb{P}(\sup_{t \in [0,T]} \|\Phi_t^n\| > \frac{n}{3}) + \mathbb{P}(\sup_{t \in [0,T]} \|\int_0^t a_n(s, Z_s^n)ds\| > \frac{n}{3})$$

$$+ \mathbb{P}(\sup_{t \in [0,T]} \|\int_0^t \int_E f_n(s, x, Z_s^n) ds q(ds, dx)\| > \frac{n}{3})$$

$$\leq \left(\frac{3}{n}\right)^2 \mathbb{E}[\sup_{t \in [0,T]} \|\Phi_t\|^2] + \left(\frac{3}{n}\right)^2 \mathbb{E}\left[\|\int_0^T a_n(s, Z_s^n) ds\|^2\right]$$

$$+ \left(\frac{3}{n}\right)^2 \mathbb{E}\left[\|\int_0^T \int_E f_n(s, x, Z_s^n) q(ds, dx)\|^2\right]$$

$$\leq \left(\frac{3}{n}\right)^2 \mathbb{E}[\sup_{t \in [0,T]} \|\Phi_t\|^2] + \left(\frac{3}{n}\right)^2 T \mathbb{E}\left[\int_0^T \|a_n(s, Z_s^n) ds\|^2\right]$$

$$\times \left(\frac{3}{n}\right)^2 K^2 \mathbb{E}\left[\int_0^T \int_E \|f_n(s, x, Z_s^n)\|^2 ds \beta(dx)\right]$$

$$\leq \left(\frac{3}{n}\right)^2 \mathbb{E}[\sup_{t \in [0,T]} \|\Phi_t\|^2] + \left(\frac{3}{n}\right)^2 M(T + K^2) \int_0^T (\|Z_s^n\|^2 + 1) ds$$

$$\leq \left(\frac{3}{n}\right)^2 \mathbb{E}[\sup_{t \in [0,T]} \|\Phi_t\|^2]$$

$$+ \left(\frac{3}{n}\right)^2 M(T + K^2) T + \left(\frac{3}{n}\right)^2 M(T + K^2) \int_0^T \|Z_s^n\|^2 ds$$

$$\leq \left(\frac{3}{n}\right)^2 \mathbb{E}[\sup_{t \in [0,T]} \|\Phi_t\|^2] + \left(\frac{3}{n}\right)^2 M(T + K^2) T$$

$$+ \left(16 \mathbb{E}[\|\Phi_t\|^2] + 16M(1 + K^2) T\right) T e^{16M(1+K^2)T}.$$

This gives

$$\mathbb{P}(\cap_n \{\tau_n < T\}) = \lim_n \mathbb{P}(\{\tau_n < T\}) = 0.$$

Thus we get the solution which is in S_T^2. □

4.4 Continuous Dependence on Initial Data, Coefficients and the Markov Property

Theorem 4.4.1 *Let F be a separable Banach space of M-type 2. Let T > 0. Assume $\Phi|_{[0,T]} \in S_T^2$ for all $T \geq 0$ and that, like $a(t, x)$ and $f(t, x, z)$, $a^n(t, x)$ and $f^n(t, x, z)$ are $(\mathcal{F}_t)$-adapted functions satisfying conditions (B1) and (B2). Assume also that $a(t, x)$ and $f(t, x, z)$ satisfy conditions (B3') and (B4) and*

(C1) *for each $c > 0$ there exists a constant $L(c) > 0$ such that*

$$T\|a^n(t,z) - a^n(t,y)\|_F^2 + \int_0^T \int_E \|f^n(t,x,z) - f_n(t,x,y)\|_F^2 \beta(dx)$$

$$\leq L(c)\|z - y\|^2 \quad \text{for all } t \in [0,T], \, z, y \in F, \, \|z\| < c, \, \|y\| < c, \, n \in \mathbb{N} \quad P - a.e.$$

(C2) *There exists a constant $M > 0$ such that*

$$\|a^n(t,z)\|_F^2 + \int_E \|f^n(t,x,z)\|_F^2 \beta(dx) \leq M(\|z\|^2 + 1) \quad \text{for all } t \in [0,T], z \in F, n \in \mathbb{N} \quad \mathbb{P} - a.e.$$

(C3) $\sup_n \mathbb{E}[\sup_{t \in [0,T]} \|\phi_t^n\|^2] < \infty$
(C4) *for all $t \in [0,T], z \in F$*

$$\|\phi_t^n - \phi_t\| + \|a^n(t,z) - a(t,z)\|_F^2$$

$$+ \int_0^T \int_E \|f^n(t,x,z) - f(t,x,z)\|_F^2 \beta(dx)dt \to 0 \quad \text{in probability as } n \to \infty.$$

Let $(Z_t^n)_{n \in \mathbb{K}}$ be the solutions of (4.2.1) with coefficients $a^n(t,x)$, $f^n(t,x,z)$ and ϕ_t^n, respectively, then for each $t \in [0,T]$, $Z_t^n \to Z_t$ in probability as $n \to \infty$.

Proof Under the given assumptions $(Z_t^n)_{t \in [0,T]}$ exists and is unique. By Doob's inequality (Theorem 2.3.5), similarly to the proof of (4.3.5), it can be shown that there is a constant C such that

$$\mathbb{E}[\sup_{t \in [0,T]} \|Z_t^n\|^2] \leq e^{CT} \mathbb{E}[\sup_{t \in [0,T]} \|\phi_t^n\|^2]. \qquad (4.4.1)$$

Let us define

$$\psi_n^N(t) := \begin{cases} 1 & \text{if } \|\Phi_s^n\| + \|\phi_s\| + \|Z_s^n\| + \|Z_s\| \leq N, \, \forall s \in [0,T], \\ 0 & \text{if } \|\Phi_s^n\| + \|\phi_s\| + \|Z_s^n\| + \|Z_s\| > N, \, \text{for some } s \in [0,T], \end{cases}$$

$$(Z_t^n - Z_t)\psi_n^N(t) = (\Phi_t^n - \Phi_t)\psi_n^N(t) + \psi_n^N(t)\{\int_0^t (a^n(s,Z_s) - a(s,Z_s))ds$$

$$+ \int_0^t \int_{E \backslash \{0\}} (f_n(s,x,Z_s) - f(s,x,Z_s))q(dsdu)\}.$$

For $s \leq t$, $\psi_n^N(t) \leq \psi_n^N(s)$, $s, t \in [0,T]$. It follows from (C2) that

$$\mathbb{E}[\|Z_t^n - Z_t\|^2 \psi_n^N(t)] \leq \mathbb{E}[\alpha_n^N(t)] + C \int_0^t \mathbb{E}[\|Z_s^n - Z_s\|^2 \psi_n^N(s)],$$

where $\alpha_n^N(t) = (\Phi_t^n - \Phi_t)\psi_n^N(t)$.
$\alpha_n^N(t) \to 0$ in probability uniformly for $t \in [0,T]$. From $\alpha_n^N(t) \leq 4N^p$ and the dominated convergence theorem it follows that

$$\mathbb{E}[\alpha_n^N(t)] \to 0 \quad \text{uniformly in} \quad t \in [0, T].$$

By Gronwall's Lemma

$$\mathbb{E}[\|Z_t^n - Z_t\|^2 \psi_n^N(t)] \to 0 \quad \text{as} \quad n \to \infty.$$

Let $\epsilon > 0$, then

$$\mathbb{P}(\|Z_t^n - Z_t\| > \epsilon) \leq \frac{1}{\epsilon} \mathbb{E}[\|Z_t^n - Z_t\| \psi_n^N(t)] + \mathbb{P}(\psi_n^N(t) = 0).$$

Using (C3) it is easy to check

$$\mathbb{P}(\psi_n^N(t) = 0) \to 0 \quad \text{as} \quad N \to \infty. \tag{4.4.2}$$

Hence

$$\lim_{N \to \infty} \limsup_n \mathbb{P}(\|Z_t^n - Z_t\| > \epsilon) \leq \frac{1}{\epsilon^2} \lim_{N \to \infty} \limsup_n \mathbb{E}[\|Z_t^n - Z_t\|^2 \psi_n^N(t)]$$
$$+ \lim_{N \to \infty} \limsup_n \mathbb{P}(\psi_n^N(t) = 0),$$

completing the proof. $\qquad\square$

Exercise Prove (4.4.2).

Now let us consider $a(t, z) = a(z)$, $\Phi(t, \omega) = \Phi(t)$ and $f(t, x, z) = f(x, z)$ for all $t \geq 0$. Then, as in the classical one-dimensional case, the solution is homogenous [79].

Theorem 4.4.2 *Let $a(t, z) = a(z), f(t, x, z) = f(x, z)$ and $\Phi(t) = z \in F$ for all $t \geq 0$, let F be a Banach space of M-type 2 and assume (B1)–(B4) are satisfied. Then the solution of (4.2.1) is Markov.*

Proof We follow the classical method. We denote by $Z_t(s; u)$ the solution of (4.2.1) starting with $Z_s = u$ at time s, i.e.

$$Z_t(s; u) = u + \int_s^t a(r, Z_r(s; u))dr + \int_s^t \int_E f(r, x, Z_r(s; u))q(dr, dx)$$

and define for a bounded measurable function Ψ on F and for $u \in F$

$$P_{s,t}(\Psi)(u) = \mathbb{E}[\Psi(Z_t(s; u))].$$

We have to prove the Markov property, i.e. for any $v \leq s \leq t \leq T$

$$\mathbb{E}[\Psi(Z_t(v; \xi))/\mathcal{F}_s^Z] = (P_{s,t})(\Psi)(Z_s(v; \xi)) \quad \text{for any} \quad \Psi \in B_b(F). \tag{4.4.3}$$

Here $B_b(F)$ is the space of bounded measurable function on F.

As $\mathcal{F}_s^Z \subset \mathcal{F}_s$ it is sufficient to prove

$$\mathbb{E}[\Psi(Z_t(v; \xi))/\mathcal{F}_s] = P_{s,t}(\Psi)(Z_s(v; \xi)) \qquad (4.4.4)$$

for any $\mathcal{F}_v$-adapted random variable ξ, any $v \leq s \leq t \leq T$ and any Ψ bounded measurable function on F.

From uniqueness of the solution of (4.2.1) it follows that

$$Z_t(v; \xi)(\omega) = Z_t(s; Z_s(v; \xi)(\omega))(\omega) \quad P - a.s. \qquad (4.4.5)$$

Let

$$\eta(\omega) := Z_s(v; \xi)(\omega). \qquad (4.4.6)$$

Then from (4.4.5) it follows that (4.4.4) can be written as

$$\mathbb{E}[\Psi(Z_t(s; \eta))/\mathcal{F}_s] = P_{s,t}(\Psi)(\eta). \qquad (4.4.7)$$

We will prove (4.4.7) for all $\sigma(Z_s(v; x))$-measurable random variables η with $\mathbb{E}[\|\eta\|^2 < \infty$. Note that, due to Theorem 4.2.2, $E\|Z_s(v; \xi)\|^2 < \infty$.

If $\eta = x$ then $Z_t(s; x)$ is independent of $\mathcal{F}_s$ by definition of the Lévy process, associated cPrm and Itô integral. It follows that

$$\mathbb{E}[\Psi(Z_t(s; x)/\mathcal{F}_s] = \mathbb{E}[\Psi(Z_t(s; x))] = P_{s,t}(\Psi(x)), \qquad (4.4.8)$$

so that (4.4.7) holds for this particular case.

Now we prove (4.4.7) for the case where

$$\eta(\omega) := \sum_1^n a_j \mathbf{1}_{A_j}(Z_s(v; \xi)) \qquad (4.4.9)$$

with $\{A_j, j = 1, \ldots, n\}$ a partition of F and $a_1, \ldots, a_n \in F$. In this case

$$Z_t(s; \eta(\omega))(\omega) = \sum_1^n Z_t(s; a_j) \mathbf{1}_{A_j}(Z_s(v; \xi)) \quad P - a.s.,$$

$$\Psi(Z_t(s; \eta(\omega))(\omega)) = \sum_1^n \Psi(Z_t(s; a_j)) \mathbf{1}_{A_j}(Z_s(v; \xi)) \quad P - a.s.,$$

and

$$\mathbb{E}[\Psi(Z_t(s;\eta)/\mathcal{F}_s] = \mathbb{E}[\sum_1^n \Psi(Z_t(s;a_j))\mathbf{1}_{A_j}(Z_s(v;\xi))/\mathcal{F}_s] \qquad (4.4.10)$$

$$= \sum_1^n P_{s,t}(\Psi)(a_j)\mathbf{1}_{A_j}Z_s(v,\xi) = P_{s,t}(\Psi)(\eta), \qquad (4.4.11)$$

where we used that $\Psi(Z_t(s;a_j))$ are independent of $\mathcal{F}_s$ and $\mathbf{1}_{A_j}(Z_s(v;\xi))$ are $\mathcal{F}_s$-measurable.

If $\mathbb{E}[\|\eta\|^2] < \infty$ then there exists a sequence of η_n of the form (5.4.7) such that $\mathbb{E}[\|\eta_n - \eta\|^2 \to 0]$ and by Theorem 4.4.1, using a subsequence and the fact that $\psi(Z_t(s;\eta))$ is bounded, we get the result.   □

Exercise Take a general η and prove the result, by approximating with $\eta_k = \eta \wedge k$.

4.5 Remarks and Related Literature

The material of Sects. 4.2–4.4 is taken from [65] and Sect. 4.1 gives generalizations of results in [36] to the Banach space case (see also [35]).

Note that in this chapter the geometry of Banach spaces is only used in the definition of the Itô integral. In view of condition (3.1.4), the results are valid in any Banach space for which the driving jump process satisfies (3.1.4).

Chapter 5
Stochastic Partial Differential Equations in Hilbert Spaces

In this chapter we study partial differential equations. It is well known [83] that finite dimensional partial differential equations lead to infinite-dimensional ordinary differential equations in the deterministic case involving unbounded operators. The solutions of these can be studied by semigroup methods. However, one has to distinguish between classical solutions and so-called mild solutions. In the stochastic case involving Gaussian noise they are studied in the book [34]. In order to keep our presentation self-contained, we describe in the next section the basic theory of semigroups and how it is used in solving deterministic partial differential equations. This material is taken from [83], where the complete proofs can be found.

5.1 Elements of Semigroup Theory

Let $(E, \| \cdot \|_E)$, $(F, \| \cdot \|_F)$ be Banach spaces and $\mathbf{L}(E/F)$ be the space of bounded linear operators from E to F. It is known that $\mathbf{L}(E/F)$ is a Banach space, when equipped with the norm

$$\|T\|_{\mathbf{L}(E/F)} = \sup_{\|x\|_E=1} \|Tx\|_F, \quad T \in \mathbf{L}(E/F). \qquad (5.1.1)$$

We denote by $\mathbf{L}(F) = \mathbf{L}(F/F)$ and by $Id \in \mathbf{L}(F)$ the identity operator.

For $T \in \mathbf{L}(E/F)$, we recall $T^\star \in \mathbf{L}(F^\star, E^\star)$ defined by $\langle x, T^\star y^\star \rangle = \langle Tx, y^\star \rangle$, $x \in E$, $y^\star \in F^\star$, is the adjoint operator of T. If $E = F = H$ is a Hilbert space, the operator T is said to be symmetric if $T = T^\star$, and is non-negative if $\forall h \in H$, $\langle Th, h \rangle \geq 0$.

Definition 5.1.1 A family $\{(S_t), t \geq 0\} \subset \mathbf{L}(E)$ is called a strongly continuous semigroup (C_0-semigroup for short) if the following properties hold:

- $S_0 = \mathrm{Id}$;
- *(semigroup property)* $S_{s+t} = S_s S_t$ for all $s, t \geq 0$;
- *(strong continuity property)* $\lim_{t \to 0} S_t x = x$ for all $x \in E$.

© Springer International Publishing Switzerland 2015
V. Mandrekar and B. Rüdiger, *Stochastic Integration in Banach Spaces*,
Probability Theory and Stochastic Modelling 73, DOI 10.1007/978-3-319-12853-5_5

Let $\{S_t\}$ be a C_0-semigroup in a Banach space E. Then there exist constants $\alpha \geq 0$ and $M \geq 1$ such that

$$\|S_t\|_{\mathbf{L}(E)} \leq M e^{\alpha t} \quad t \geq 0. \tag{5.1.2}$$

If $M = 1$, then $\{S_t\}$ is called a "pseudo-contraction semigroup". If $\alpha = 0$ then $\{S_t\}$ is said to be "uniformly bounded" and if $\alpha = 0$ and $M = 1$, then $\{S_t\}$ is called a "contraction semigroup". If for every $x \in E$, $t \to S_t x$ is differentiable for $t > 0$, then $\{S_t\}$ is called a "differentiable semigroup".

Note that for a C_0-semigroup, $t \to S_t x$ is continuous for $x \in E$.

Definition 5.1.2 Let $\{S_t\}$ be a C_0-semigroup on E. The linear operator A with domain

$$\mathcal{D}(A) := \{x \in E, \lim_{t \to 0^+} \frac{S_t x - x}{t} \; exists\}$$

defined by

$$Ax = \lim_{t \to 0^+} \frac{S_t x - x}{t}$$

is called the infinitesimal generator of $\{S_t\}$.

We call $\{S_t\}$ "uniformly continuous" if $\lim_{t \to 0^+} \|S_t - I\|_{\mathbf{L}(E)} = 0$. In this case $\{S_t\}$ is uniformly continuous iff A is bounded and

$$S_t = e^{tA} = \sum_{n=0}^{\infty} \frac{(tA)^n}{n!}$$

with the convergence in norm for every $t \geq 0$.

Theorem 5.1.3 *Let A be the infinitesimal generator of a C_0-semigroup $\{S_t\}$ on E, then*

(1) *For $x \in E$*

$$\lim_{h \to 0} \frac{1}{h} \int_t^{t+h} S_s x \, ds = S_t x.$$

(2) *For $x \in \mathcal{D}(A)$, $S_t x \in \mathcal{D}(A)$ and*

$$\frac{d}{dt} S_t x = A S_t x = S_t A x.$$

(3) *For $x \in E$, $\int_0^t S_s x ds \in \mathcal{D}(A)$ and*

$$A \int_0^t S_s x ds = S_t x - x.$$

(4) *If $\{S_t\}$ is differentiable then for $n = 1, 2, \ldots$, $S_t : E \to \mathcal{D}(A^n)$ and $S_t^n := A^n S_t \in \mathbf{L}(E)$.*

(5) *For $x \in \mathcal{D}(A)$*

$$S_u x - S_t x = \int_t^u S_s A x ds = \int_t^u A S_t x ds.$$

(6) *$\mathcal{D}(A)$ is dense in E and A is a closed operator.*

Furthermore $\cap_n \mathcal{D}(A^n)$ is dense in E, and if E is reflexive, then the adjoint semigroup $\{S_t^*\}$ of $\{S_t\}$ is a C_0-semigroup with infinitesimal generator A^*, the adjoint of A.

We shall be dealing with $E = H$, a real separable Hilbert space. In this case, for $h \in H$, we define the graph norm

$$\|h\|_{\mathcal{D}(A)} := (\|h\|_H^2 + \|Ah\|_H^2)^{1/2}.$$

Then $(\mathcal{D}(A), \|\cdot\|_{\mathcal{D}(A)})$ is a real separable Hilbert space.

Exercise Let A be a closed linear operator on a real separable Hilbert space. Prove that $(\mathcal{D}(A), \|\cdot\|_{\mathcal{D}(A)})$ is a real separable Hilbert space.

Let $\mathcal{B}(H)$ be the Borel σ-field on H. Then $\mathcal{D}(A) \in \mathcal{B}(H)$ and

$$A : (\mathcal{D}(A), \mathcal{B}(H)|_{\mathcal{D}(A)}) \to (H, \mathcal{B}(H)).$$

Consequently, $\mathcal{B}(H)|_{\mathcal{D}(A)}$ coincides with the Borel σ-field on the Hilbert space $(\mathcal{D}(A), \|\cdot\|_{\mathcal{D}(A)})$.

Measurability of $\mathcal{D}(A)$-valued functions can be understood with respect to either of the two σ-fields.

Theorem 5.1.4 *Let $f : [0, T] \to \mathcal{D}(A)$ be measurable and let $\int_0^t \|f(s)\|_{\mathcal{D}(A)} ds < \infty$. Then*

$$\int_0^t f(s)ds \in \mathcal{D}(A) \quad and \quad \int_0^t Af(s)ds = A \int_0^t f(s)ds.$$

Now we introduce the concept of the resolvent of A, which is needed for Yosida approximation.

Definition 5.1.5 The resolvent set $\rho(A)$ of a closed linear operator A on a Banach space E is the set of all complex numbers λ for which $\lambda I - A$ has a bounded inverse $R(\lambda, A) := (\lambda I - A)^{-1} \in \mathbf{L}(E)$. The family of linear operators $R(\lambda, A)$, $\lambda \in \rho(A)$, is called the resolvent of A.

We note that $R(\lambda, A) : E \to \mathcal{D}(A)$ is one-to-one, i.e.

$$(\lambda I - A)R(\lambda, A)x = x, \quad x \in E$$

$$\text{and} \quad R(\lambda, A)(\lambda I - A)x = x, \quad x \in \mathcal{D}(A),$$

$$\text{giving} \quad AR(\lambda, A)x = R(\lambda, A)Ax, \quad x \in \mathcal{D}(A).$$

Exercise Show that $R(\lambda_1, A)R(\lambda_2, A) = R(\lambda_2, A)R(\lambda_1, A)$ for $\lambda_1, \lambda_2 \in \mathcal{D}(A)$.

Lemma 5.1.6 *Let $\{S_t\}$ be a C_0-semigroup with infinitesimal generator A. Let*

$$\alpha_0 := \lim_{t \to \infty} t^{-1} ln(\|S_t\|_{L(E)}),$$

then any real number $\lambda > \alpha_0$ belongs to the resolvent set $\rho(A)$ and

$$R(\lambda, A)x = \int_0^\infty e^{-\lambda t} S_t x dt \quad x \in E.$$

In addition, for $x \in E$

$$\lim_{\lambda \to \infty} \|\lambda R(\lambda, A)x - x\|_E = 0.$$

Theorem 5.1.7 (Hille–Yosida Theorem) *Let $A : \mathcal{D}(A) \subset E \to E$ be a linear operator on a Banach space E. Necessary and sufficient conditions for A to generate a C_0-semigroup are*

(1) *A is closed and $\overline{\mathcal{D}(A)} = E$.*
(2) *There exist $\alpha, M \in \mathbb{R}$ such that for $\lambda > \alpha$, $\lambda \in \rho(A)$*

$$\|R(\lambda, A)^r\|_{L(E)} \le M(\lambda - \alpha)^{-r}, \quad r = 1, 2, \ldots$$

In this case $\|S_t\|_{L(E)} \le Me^{\alpha t}, t \ge 0$.

For $\lambda \in \rho(A)$, consider the family of operators

$$R_\lambda := \lambda R(\lambda, A).$$

Since the range $\mathcal{R}(R(\lambda, A))$ of $R(\lambda, A)$ is such that $\mathcal{R}(R(\lambda, A)) \subset \mathcal{D}(A)$, we define the "Yosida approximation" of A by

$$A_\lambda x = AR_\lambda x, \quad x \in E.$$

Exercise Use $\lambda(\lambda I - A)R(\lambda, A) = \lambda I$ to prove

$$A_\lambda x = \lambda^2 R(\lambda, A) - \lambda I.$$

From the exercise, $A_\lambda \in \mathbf{L}(E)$. Denote by S_t^λ the uniformly continuous semigroup

$$S_t^\lambda x = e^{tA_\lambda}x, \quad x \in E.$$

Using the commutativity of the resolvent, we get $A_{\lambda_1} A_{\lambda_2} = A_{\lambda_2} A_{\lambda_1}$, and clearly

$$A_\lambda S_t^\lambda = S_t^\lambda A_\lambda.$$

Theorem 5.1.8 (Yosida approximation) *Let A be an infinitesimal generator of a C_0-semigroup $\{S_t\}$ on a Banach space E. Then*

(a) $\lim_{\lambda \to \infty} R_\lambda x = x, \quad x \in E$.
(b) $A_\lambda x = Ax, \quad for \ x \in \mathcal{D}(A)$.
(c) $\lim_{\lambda \to \infty} S_t^\lambda x = S_t x, \quad x \in E$.

The convergence in (c) is uniform on compact subsets of $\mathbb{R}_+$ and

$$\|S_t^\lambda\|_{\mathbf{L}(E)} \leq M \exp\left(\frac{t \wedge \alpha}{(\lambda - \alpha)}\right)$$

with constants M and α as in the Hille–Yosida Theorem.

We conclude this section by introducing the concept of a "mild" solution. Let us look at the deterministic problem

$$\frac{du(t)}{dt} = Au(t), \ u(0) = x, \quad x \in H.$$

Here H is a real separable Hilbert space and A is an unbounded operator generating a C_0-semigroup.

A classical solution $u : [0, T] \to H$ of the above equation will require a solution to be continuously differentiable and $u(t) \in \mathcal{D}(A)$. However,

$$u^x(t) = S_t x, \quad t \geq 0$$

is considered as a solution to the equation [83, Capt. 4]. For $x \notin \mathcal{D}(A)$, it is not a classical solution. Such a solution is called a "mild solution".

In fact, one can consider the non-homogeneous equation

$$\frac{du(t)}{dt} = Au(t) + f(t, u(t)), \ u(0) = x, \quad x \in H.$$

Then for Bochner integrable $f \in L^1([0, T], H)$, one can consider the integral equation

$$u^x(t) = S_t x + \int_0^t S_{t-s} f(s, u(s)) ds.$$

A solution of this equation is called a "mild solution" if $u \in C([0, T], H)$.

We will consider mild solutions of stochastic partial differential equations (SPDEs) with Poisson noise. Note that the stochastic integral $\int_0^t S_{t-s} f(s, x) q(ds, dx)$, which appears in such SPDEs, is in general not a martingale. However, as for Doob's inequality, the following lemma holds.

Lemma 5.1.9 *Assume $(S_t)_{t \geq 0}$ is pseudo-contractive. Let $q(ds, dx)$ be a compensated Poisson random measure on $\mathbb{R}_+ \times E$ associated to a Poisson random measure N with compensator $dt \otimes \beta(dx)$. For each $T \geq 0$ the following statements are valid:*

1. *There exists a constant $C > 0$ such that for each $f \in \mathcal{L}^2_{T,\beta}(E, H)$ we have*

$$\mathbb{E}\left[\sup_{t \in [0,T]} \left\| \int_0^t S_{t-s} f(s, x) q(ds, dx) \right\|^2 \right] \leq Ce^{2\alpha T} \mathbb{E}\left[\int_0^T \int_E \|f(s, x)\|^2 \beta(dx) ds \right].$$

$$(5.1.3)$$

2. *For all $f \in \mathcal{L}^2_{T,\beta}(E, H_0)$ and all $\epsilon > 0$ we have*

$$\mathbb{P}\left[\sup_{t \in [0,T]} \left\| \int_0^t S_{t-s} f(s, x) q(ds, dx) \right\| > \epsilon \right]$$

$$\leq \frac{4e^{2\alpha T}}{\epsilon^2} \mathbb{E}\left[\int_0^T \int_E \|f(s, x)\|^2 \beta(dx) ds \right], \qquad (5.1.4)$$

where $\int_0^t S_{t-s} f(s, x) q(ds, dx)$ is well defined, if the r.h.s. is finite. $\int_0^t S_{t-s} f(s, x) q(ds, dx)$ is càdlàg.

Proof Let M be the martingale

$$M_t = \int_0^t \int_E f(s, x) q(ds, dx), \quad t \in [0, T].$$

By Theorem 3.6.5 we have

$$\int_0^t S_{t-s} f(s, x) q(ds, dx) = \int_0^t S_{t-s} dM_s, \quad t \in [0, T].$$

Using [44, Theorem 3'.22'] we obtain

$$\mathbb{E}\left[\sup_{t \in [0,T]} \left\| \int_0^t S_{t-s} f(s, x) q(ds, dx) \right\|^2 \right] \leq e^{2\alpha T} (3 + \sqrt{10})^2 \mathbb{E}[\langle M, M \rangle_T]$$

$$= e^{2\alpha T} (3 + \sqrt{10})^2 \mathbb{E}\left[\int_0^T \int_E \|f(s, x)\|^2 \beta(dx) ds \right],$$

proving (5.1.3) with $C = (3 + \sqrt{10})^2$, and using [44, Theorem. $5'.16'$] we obtain

$$\mathbb{P}\left[\sup_{t \in [0,T]} \left\| \int_0^t S_{t-s} f(s, x) q(ds, dx) \right\| > \epsilon \right] \leq \frac{4e^{2\alpha T}}{\epsilon^2} \mathbb{E}[\langle M, M \rangle_T]$$

$$= \frac{4e^{2\alpha T}}{\epsilon^2} \mathbb{E}\left[\int_0^T \int_E \| f(s, x) \|^2 \beta(dx) ds \right],$$

proving (5.1.4).

Let us show that $\int_0^t S_{t-s} f(s, x) q(ds, dx)$ is càdlàg. There is a sequence of simple functions $\{f_n\}_{n \in \mathbb{N}}$ such that $f_n \to f$ in $\mathcal{L}_2^\beta(H)$, i.e.

$$\lim_{n \to \infty} \int_0^T \int_H \mathbb{E}[\| f_n(t, u) - f(t, u) \|^2] \, dt \, \beta(du) = 0.$$

Let

$$Y_t^n := \int_0^t \int_H S_{t-s} f_n q(ds, du) = \int_0^t S_{t-s} dM_s^n,$$

$$M_s^n := \int_0^t \int_H f_n q(ds, du). \tag{5.1.5}$$

As $S_{t-s} f_n(s, u, \omega)$ belongs to the set $\Sigma(H)$ of simple functions, Y_t^n is a martingale and is càdlàg.

It follows that

$$\mathbb{P}\left(\sup_{0 \leq t \leq T} \| Y_t^n - Y_t^m \| > \epsilon \right) \leq 4 \frac{e^{2\alpha T}}{\epsilon^2} \mathbb{E}[\langle M_n - M_m \rangle_T]$$

$$\leq 4 \frac{e^{2\alpha T}}{\epsilon^2} \int_0^T \int_H \mathbb{E}[\| f_n(t, u) - f_m(t, u) \|^2] \, dt \, \beta(du).$$

By the Borel–Cantelli Lemma and $f_n \to f$ in $\mathcal{L}_2^\beta(H)$ there is a subsequence $\{Y_t^{n_k}(\omega)\}_{k \in \mathbb{N}}$ such that

$$\lim_{k \to \infty} \sup_{0 \leq t \leq T} \| Y_t^{n_k}(\omega) - Y_t^{n_{k+1}}(\omega) \| = 0 \quad \mathbb{P} - a.s.$$

It follows that

$$Y_t(\omega) = \lim_{k \to \infty} Y_t^{n_k}(\omega) \quad \text{uniformly in } [0, T], \quad \mathbb{P} - a.s.$$

We see that Y_t is càdlàg, since $Y_t^{n_k}$ is càdlàg. $\qquad \square$

5.2 Existence and Uniqueness of Solutions of SPDEs Under Adapted Lipschitz Conditions

We shall study in this section càdlàg solutions to stochastic partial differential equations with non-Gaussian noise. As stated in Sect. 5.1, we shall treat these equations as ordinary stochastic differential equations in an infinite-dimensional space involving unbounded operators. Let us now assume that H is a separable Hilbert space and let A be a (generally unbounded) linear operator on the domain $\mathcal{D}(A) \subset H$. Assume that A is an infinitesimal generator of a pseudo-contraction semigroup $\{S_t\}_{t \geq 0}$ on H to H.

We want to study the existence and uniqueness of mild solutions of the stochastic differential equation on the interval $[0, T]$

$$\begin{cases} dZ_t = (AZ_t + a(t, Z))dt + \int_H f(t, u, Z)q(dt, du) \\ Z_0 = Z_0(\omega), \end{cases} \qquad (5.2.1)$$

where $a(\cdot, z)$, $f(\cdot, u, z)$ are, for fixed $z \in H$, $u \in H$, functions on $D(\mathbb{R}_+, H)$ and Z_t is Z evaluated in t. In other words, we look at the solution of the integral equation

$$Z_t = S_t Z_0 + \int_0^t S_{t-s} a(s, Z)ds + \int_0^t \int_H S_{t-s} f(s, u, Z)q(ds, du), \qquad (5.2.2)$$

where the integrals on the r.h.s. are well defined.

As in Chap. 4, we assume that with $\Omega = D(\mathbb{R}_+, H)$, $\mathcal{F}_t$ is a σ-algebra generated by cylinder sets of Ω with base on $[0, t]$. Let us assume throughout that A is an infinitesimal generator of a pseudo-contraction C_0-semigroup. Let

$$a : \mathbb{R}_+ \times D(\mathbb{R}_+, H) \to H, \quad f : \mathbb{R}_+ \times H \times D(\mathbb{R}_+, H) \to H$$

be functions and $\|z\|_\infty := \sup_{0 \leq t \leq T} \|z(t)\|_H$, for $T < \infty$.

(a) $f(t, u, z)$ is jointly measurable and, for each $t \in \mathbb{R}_+$, $u \in H$, $f(t, u, \cdot)$ is $\mathcal{F}_t$-adapted.
(b) $a(t, z)$ is jointly measurable and, for each $t \in \mathbb{R}_+$, $a(t, \cdot)$ is $\mathcal{F}_t$-adapted.

If we consider the map $\theta_t : D(\mathbb{R}_+, H) \to D(\mathbb{R}_+, H)$ defined by

$$\theta_t(z)(s) = z_s \quad \text{if} \ \ 0 \leq s \leq t$$
$$= z_t \quad \text{if} \ \ t \leq s$$

then $f(t, u, z) = f(t, u, \theta_t(z))$ and $a(t, z) = a(t, \theta_t(z))$.

(c) There exists a constant $l > 0$ such that, for fixed $t_1, t_2 \in [0, T]$,

$$\int_{t_1}^{t_2} \int_H \|f(t, u, z)\|^2 dt \beta(du) + \int_{t_1}^{t_2} \|a(t, z)\|_H^2 dt$$
$$\leq l \int_{t_1}^{t_2} (1 + \|\theta_t(z)\|_\infty^2) dt \quad \mathbb{P} - a.s.$$

(d) There exists a constant $K > 0$ such that, for fixed $t_1, t_2 \in [0, T]$ and $z, y \in D(\mathbb{R}_+, H)$,

$$\int_{t_1}^{t_2} \int_H \|f(t, u, z) - f(t, u, y)\|^2 dt \beta(du) + \int_{t_1}^{t_2} \|a(t, z) - a(t, y)\|_H^2 dt$$
$$\leq K \int_{t_1}^{t_2} \|\theta_t(z) - \theta_t(y)\|_\infty^2 dt \quad \mathbb{P} - a.s.$$

Let, for $Z \in D(\mathbb{R}_+; H)$,

$$I(t, Z) := \int_0^t S_{t-s} a(s, Z) ds + \int_0^t \int_{H \setminus \{0\}} S_{t-s} f(s, u, Z) q(ds, du), \quad t \in [0, T].$$
$$(5.2.3)$$

Theorem 5.2.1 *Assume (a), (b) and (c). There exists a constant $C_{l,T,\alpha}$ such that for any $\mathcal{F}_t$-stopping time τ*

$$\mathbb{E}[\sup_{0 \leq s \leq t \wedge \tau} \|I(s, Z)\|_H^2] \leq C_{l,T,\alpha}(t + \int_0^t \mathbb{E}[\sup_{0 \leq v \leq s \wedge \tau} \|Z_v\|^2] ds), \quad t \in [0, T].$$
$$(5.2.4)$$

Proof

$$\sup_{0 \leq s \leq t \wedge \tau} \|I(s, Z)\|_H^2 \leq 2 \sup_{0 \leq s \leq t \wedge \tau} \| \int_0^s S_{s-v} a(v, Z) dv \|_H^2$$
$$+ 2 \sup_{0 \leq s \leq t \wedge \tau} \| \int_0^s \int_H S_{t-v} f(v, u, Z) q(dv, du) \|_H^2$$
$$(5.2.5)$$

(where we used the inequality $\|x + y\|^2 \leq 2\|x\|^2 + 2\|y\|^2$, valid for any $x, y \in H$). Using the bound on S_t and condition (c) we obtain

$$\mathbb{E}[\sup_{0 \leq s \leq t \wedge \tau} \| \int_0^s S_{s-v} a(v, Z) dv \|_H^2]$$
$$\leq \mathbb{E}[\sup_{0 \leq s \leq t} (le^{\alpha t} \int_0^s (1 + \|\theta_v(Z)\|_\infty) dv)^2]$$
$$\leq 2e^{2\alpha T} l^2 \{t^2 + t\mathbb{E}[\int_0^{s \wedge \tau} \|\theta_v(Z)\|_\infty^2 dv]\}.$$

Moreover, using Theorem 3 of [44] and (c) we get

$$\mathbb{E}[\sup_{0\leq s\leq t\wedge\tau}\|\int_0^s\int_H S_{t-v}f(v,u,Z)q(dv,du)\|_H^2]$$

$$\leq 2e^{2\alpha T}l^2(3+\sqrt{10})^2\{t^2+t\mathbb{E}[\int_0^{s\wedge\tau}\|\theta_v(Z)\|_\infty^2 dv]\},$$

$$\mathbb{E}[\sup_{0\leq s\leq t\wedge\tau}\|I(s,Z)\|_H^2]\leq 4e^{2\alpha T}l^2(1+(3+\sqrt{10})^2)$$

$$\{t^2+t\mathbb{E}[\int_0^{s\wedge\tau}\|\theta_v(Z)\|_\infty^2 dv]\}$$

$$\leq C_{l,T,\alpha}(t+\int_0^t\mathbb{E}[\sup_{0\leq v\leq s\wedge\tau}\|Z_v\|^2]ds),$$

with $C_{l,T,\alpha}:=4Te^{2\alpha T}l^2(1+(3+\sqrt{10})^2)$.   □

Let $T>0$ and

$$\mathcal{H}_2^T:=\{\xi:=(\xi_s)_{s\in[0,T]}:\ \xi_s(\omega)\text{ is jointly measurable},$$
$$\mathcal{F}_t\text{-adapted};\mathbb{E}[\sup_{0\leq s\leq T}\|\xi_s\|_H^2]<\infty\}.$$

Let us observe that it follows from Theorem 5.2.1 that the map

$$I:\mathcal{H}_2^T\ \to\ \mathcal{H}_2^T$$
$$\xi\ \to\ I(\cdot,\xi)$$

is well defined.

Lemma 5.2.2 *Assume (a), (b), (c) and (d). The map $I:\mathcal{H}_2^T\to\mathcal{H}_2^T$ is continuous. There is a constant $C_{\alpha,K,T}$, depending on α, K and T, such that*

$$\mathbb{E}[\sup_{0\leq s\leq T}\|I(s,Z^1)-I(s,Z^2)\|_H^2]\leq C_{\alpha,K,T}\int_0^T\mathbb{E}[\sup_{0\leq s\leq T}\|Z_s^2-Z_s^1\|_H^2]ds.$$

$$(5.2.6)$$

Exercise Use (d) and the arguments as in Theorem 5.2.1 to prove Lemma 5.2.2.

Theorem 5.2.3 *Let $T>0$, $x\in H$. There is a unique solution $Z:=(Z_s)_{s\in[0,T]}$ in $\mathcal{H}_2^T$ which satisfies*

$$Z_t=S_t x+\int_0^t S_{t-s}a(s,Z)ds+\int_0^t\int_H S_{t-s}f(s,u,Z)q(ds,du).\qquad(5.2.7)$$

Proof We shall prove that the solution can be approximated in $\mathcal{H}_2^T$ by $Z^n :=$ $(Z_s^n)_{s\in[0,T]}$, for $n \to \infty$, $n \in \mathbb{N}$, where

$$Z_s^0(\omega) := S_s x \quad P - a.s.$$
$$Z_s^{n+1}(\omega) := I(s, Z^n(\omega)).$$

Note that $(Z_t^n)_{t\in[0,T]}$ is $\mathcal{F}_t$-adapted. Let

$$v_t^n := \mathbb{E}[\sup_{0\leq s\leq t} \|Z_s^{n+1} - Z_s^n\|_H^2].$$

Then from Theorem 5.2.1 it follows that there is a constant $V_{\alpha,l,T}(x)$, depending on α, l and T and the initial data x, such that

$$v_t^0 \leq \mathbb{E}[\sup_{0\leq s\leq T} \|Z_s^1 - Z_s^0\|_H^2] \leq V_{\alpha,l,T}(x).$$

Similarly as in the proof of Theorem 5.2.1, it can be proven that there is a constant $C_{\alpha,K,T}$ depending on α, K and T, such that

$$v_t^1 \leq C_{\alpha,K,T} \int_0^t \mathbb{E}[\sup_{0\leq s\leq t} \|Z_s^1 - Z_s^0\|_H^2]ds \leq \frac{T^2(C_{\alpha,K,T})^2}{2} V_{\alpha,l,T}(x).$$

In a similar way we get by induction that

$$v_t^n \leq C_{\alpha,K,T} \int_0^t v_s^{n-1}ds \leq \frac{(TC_{\alpha,K,T})^{n+1}}{(n+1)!} V_{\alpha,l,T}(x).$$

Let $\epsilon_n := \left(\frac{(TC_{\alpha,K,T})^{n+1}}{(n+1)!}\right)^{\frac{1}{3}}$. Then:

$$\mathbb{P}(\sup_{0\leq t\leq T} \|Z_t^{n+1} - Z_t^n\|^2 \geq \epsilon_n) \leq \frac{\frac{(TC_{\alpha,K,T})^{n+1}}{(n+1)!} V_{\alpha,l,T}(x)}{(\frac{(TC_{\alpha,K,T})^{n+1}}{(n+1)!})^{\frac{1}{3}}} = \epsilon_n^2 V_{\alpha,l,T}(x).$$

As $\sum_n \epsilon_n^2$ is convergent, we get that $\sum_{n=1}^{\infty} \sup_{0\leq t\leq T} \|Z_t^{n+1} - Z_t^n\|^2$ converges P-a.s. It follows that there is a process $Z := (Z_t)_{t\in[0,T]}$, $Z \in D([0,T]; H)$, such that Z^n converges to Z, as n goes to infinity, in the space $D([0,T]; H)$ (with the supremum norm), P-a.s. Moreover

$$\mathbb{E}[\sup_{0\leq t\leq T} \|Z_t - Z_t^n\|^2] = \mathbb{E}[\lim_{m\to\infty} \sup_{0\leq t\leq T} \|\sum_{k=n}^{n+m-1}(Z_t^{k+1} - Z_t^k)\|^2]$$

$$\leq \mathbb{E}[\lim_{m\to\infty} (\sum_{k=n}^{n+m-1} \sup_{0\leq t\leq T} \|Z_t^{k+1} - Z_t^k\| k\frac{1}{k})^2]$$

$$\leq \sum_{k=n}^{\infty} \mathbb{E}[\sup_{0\leq t\leq T} \|Z_t^{k+1} - Z_t^k\|^2 k^2)] \sum_{k=n}^{\infty} \frac{1}{k^2}$$

$$\leq V_{\alpha,l,T}(x)(\sum_{k=n}^{\infty} \frac{(TC_{\alpha,K,T})^{k+1} k^2}{(k+1)!})$$

$$(\sum_{k=n}^{\infty} \frac{1}{k^2}) \to 0 \quad \text{as} \quad n \to \infty,$$

where we used Schwarz's inequality. It follows that, as n goes to infinity, Z^n also converges to Z in the space $\mathcal{H}_2^T$. From Lemma 5.2.2 it follows that $(Z_t)_{0\leq t\leq T}$ solves (5.2.7). We shall prove that the solution is unique. Suppose that $(Z_t)_{0\leq t\leq T}$ and $(Y_t)_{0\leq t\leq T}$ are two solutions of (5.2.7). Let

$$\mathcal{V}_t := \mathbb{E}[\sup_{0\leq s\leq t} \|Z_s - Y_s\|_H^2].$$

Then similarly as before we get

$$\mathcal{V}_t \leq C_{\alpha,K,T} \int_0^t \mathcal{V}_s$$

and by induction

$$\mathcal{V}_t \leq \frac{(C_{\alpha,K,T}t)^n}{n!} \mathbb{E}[\sup_{0\leq s\leq T} \|Z_s - Y_s\|_H^2] \to 0 \quad \text{as} \quad n \to \infty$$

i.e. $\mathcal{V}_t = 0 \; \forall t \in [0, T]$.   □

5.3 Existence and Uniqueness of Solutions of SPDEs Under Markovian Lipschitz Conditions

Let us assume that we are given

$$a : \mathbb{R}_+ \times H \to H,$$

$$f : \mathbb{R}_+ \times H \times H \to H.$$

Assume

(A) $f(t, u, z)$ is jointly measurable,
(B) $a(t, z)$ is jointly measurable,
     and for fixed $T > 0$
(C) there is a constant $L > 0$ such that

$$T\|a(t, z) - a(t, z')\|^2 + \int_H \|f(t, u, z) - f(t, u, z')\|^2 \beta(du) \leq L\|z - z'\|^2$$

   for all $t \in [0, T]$, $z, z' \in F$,

(D) there is a constant $K > 0$ such that

$$T\|a(t, z)\|^2 + \int_H \|f(t, u, z)\|^2 \beta(du) \leq K(\|z\|^2 + 1)$$

   for all $t \in [0, T]$, $z \in F$.

We assume again that A is the infinitesimal generator of a pseudo-contraction semigroup $(S_t)_{t \in [0,T]}$. If we consider functions $a(t, z) = a(t, z_t)$ and $f(t, u, z) = f(t, u, z_t)$ on $D(\mathbb{R}_+, H)$ then our previous theorem tells us that the equation

$$Z_t = S_t Z_0 + \int_0^t S_{t-s} a(s, Z_s) ds$$

$$+ \int_0^t \int_H S_{t-s} f(s, u, Z_s) q(ds, du) \quad \mathbb{P} - a.s. \quad \forall t \in [0, T] \qquad (5.3.1)$$

has a unique solution on $\mathcal{H}_2^T$.

However, we shall now consider the SPDE on any filtered probability space $(\Omega, \mathcal{F}, (\mathcal{F}_t)_{t \geq 0}, \mathbb{P})$ (satisfying the usual conditions) and show that it has a unique càdlàg solution in S_T^2 as defined in Chap. 4.

Theorem 5.3.1 *Suppose assumptions (A)–(D) are satisfied. Then for $Z_0 \in L^2$ $(\Omega, \mathcal{F}_0, \mathbb{P}; H)$, there exists a unique mild solution in S_T^2 to (5.3.1), with initial condition Z_0, such that Z_t is $\mathcal{F}_t$-measurable.*

Proof Define the process

$$(\mathcal{S}Z)_t := S_t Z_0 + \int_0^t S_{t-s} a(s, Z_s) ds$$

$$+ \int_0^t \int_H S_{t-s} f(s, u, Z_s) q(ds, du) \quad t \in [0, T]. \qquad (5.3.2)$$

Consider

$$\mathbb{E}[\sup_{0 \le t \le T} \|(\mathcal{S}Z)_t\|^2] \le 3\mathbb{E}[\sup_{0 \le t \le T} \|S_t Z_0\|^2]$$

$$+ 3\mathbb{E}[\sup_{0 \le t \le T} \| \int_0^t \int_H S_{t-s} a(s, Z_s) ds \|^2]$$

$$+ 3\mathbb{E}[\sup_{0 \le t \le T} \| \int_0^t S_{t-s} f(s, u, Z_s) q(ds, du) \|^2].$$

Using the fact that $\|S_t\| \le e^{\alpha t}$ for $t \ge 0$, inequality (5.1.3) and (D) we get

$$\mathbb{E}[\sup_{0 \le t \le T} \|(\mathcal{S}Z)_t\|^2] \le 3e^{2\alpha T} \mathbb{E}[\|Z_0\|^2]$$

$$+ 3e^{2\alpha T} \mathbb{E}[\| \int_0^T a(s, Z_s) ds \|^2]$$

$$+ 3e^{2\alpha T} \mathbb{E}[\| \int_0^T \int_H f(s, u, Z_s) q(ds, du) \|^2]$$

$$\le 3e^{2\alpha T} \mathbb{E}[\|Z_0\|^2] + 3T e^{2\alpha T} \mathbb{E}[\int_0^T \|a(s, Z_s)\|^2 ds]$$

$$+ 3e^{2\alpha T} C \mathbb{E}[\int_0^T \int_H \|f(s, u, Z_s)\|^2 ds \beta(du)]$$

$$\le 3e^{2\alpha T} \mathbb{E}[\|Z_0\|^2] + 3e^{2\alpha T} C(K \mathbb{E}[\int_0^T \|Z_s\|^2 ds] + K)$$

where C is any fixed constant such that (5.1.3) holds and such that $C > 1$.

This shows that $\mathcal{S}$ maps S_T^2 into itself.

For $Y, Z \in S_T^2$ using again (5.1.3)

$$\mathbb{E}[\sup_{0 \le s \le t} \|(\mathcal{S}Y)_s - (\mathcal{S}Z)_s\|^2] \le 2\mathbb{E}[\sup_{0 \le s \le t} \| \int_0^s \int_H S_{s-v}(a(v, Y_v)$$

$$- a(v, Z_v)) dv \|^2]$$

$$+ 2\mathbb{E}[\sup_{0 \le s \le t} \| \int_0^s S_{s-v}(f(v, u, Y_v)$$

$$- f(v, u, Z_v)) q(dv, du) \|^2]$$

$$\le 2T e^{2\alpha T} \mathbb{E}[\int_0^t \|a(s, Y_s) - a(s, Z_s)\|^2 ds]$$

$$+ 2e^{2\alpha T} C \mathbb{E}[\int_0^T \int_H \|f(s, u, Y_s)$$

$$- f(s, u, Z_s)\|^2 ds \beta(du)].$$

Using (C) we get

$$\mathbb{E}[\sup_{0 \le s \le t} \|(\mathcal{S}Y)_s - (\mathcal{S}Z)_s\|^2] \le 2Le^{2\alpha T} C\mathbb{E}[\int_0^t \|Y_s - Z_s\|^2 ds]$$

$$\le 2Le^{2\alpha T} C\mathbb{E}[\int_0^t \sup_{0 \le v < s} \|Y_v - Z_v\|^2 ds].$$

By induction we get

$$\mathbb{E}[\sup_{0 \le s \le t} \|(\mathcal{S}Y)_s - (\mathcal{S}Z)_s\|^2] \le \frac{2^n (CL)^n e^{2\alpha n T}}{n!} \mathbb{E}[\sup_{0 \le t < T} \|Y_t - Z_t\|^2 ds].$$

Hence for some $n \in \mathbb{N}$, $\mathcal{S}$ is a contraction, yielding the conclusion by the fixed point theorem. $\qquad\square$

Corollary 5.3.2 *Let $0 < T < \infty$, and assume (A), (B), (C) and (D). Let $(Z_t^\xi)_{t \in [0,T]}$ (resp. $(Z_t^\eta)_{t \in [0,T]}$) be the solution to (5.3.1) with initial condition ξ (resp. η), then*

$$\mathbb{E}[\|Z_t^\xi - Z_t^\eta\|^2] \le C_{t,\alpha} \|\xi - \eta\|^2,$$

with constant $C_{t,\alpha}$ depending on t and α.

Exercise Prove the corollary by computing $\mathbb{E}[\|Z_t^\xi - Z_t^\eta\|^2]$.

We assume again that A is the infinitesimal generator of a pseudo-contraction semigroup $(S_t)_{t \in [0,T]}$ and conditions (A), (B), (C), (D) hold. Let

$$Z_0(\omega) = \xi \quad \mathbb{P}-a.s.$$

and let $(Z_t)_{t \in [0,T]}$ be the unique càdlàg process solving $\mathbb{P}$-a.s. (5.3.1) for every $t \in [0, T]$.

Let $\{A_n\}_{n \in \mathbb{N}}$ be the Yosida approximation to A (see Sect. 5.1). For every fixed $T > 0$, there exists a unique càdlàg process $(Z_t^n)_{t \in [0,T]}$ such that $\int_0^T \mathbb{E}[\|Z_s^n\|^2] ds < \infty$ and such that $(Z_t^n)_{t \in [0,T]}$ is a strong solution of

$$dZ_t^n = A_n Z_t^n dt + a(t, Z_t^n) dt + \int_H f(s, u, Z_s^n) q(ds, du)$$

with initial condition ξ (see Chap. 4 or [65]). Moreover, $(Z_t^n)_{t \in [0,T]}$ is also a mild solution, i.e. $\mathcal{P}$-a.s.

$$Z_t^n = S_t^n \xi + \int_0^t S_{t-s}^n a(s, Z_s^n) ds + \int_0^t \int_H S_{t-s}^n f(s, u, Z_s^n) q(ds, du) \qquad (5.3.3)$$

for every $t \in [0, T]$ and such that conclusions the conditions in Theorem 5.3.1 are satisfied. We shall prove the following result:

Theorem 5.3.3

$$\lim_{n \to \infty} \mathbb{E}[\|Z_t - Z_t^n\|^2] = 0$$

uniformly in $[0, T]$.

Proof We have

$$\mathbb{E}[\|Z_t - Z_t^n\|^2] \leq 2^3 \|S_t^n \xi - S_t \xi\|^2 \tag{5.3.4}$$
$$+ 2^3 \mathbb{E}[\| \int_0^t (S_{t-s} a(s, Z_s) - S_{t-s}^n a(s, Z_s^n)) ds\|^2]$$
$$+ 2^3 \mathbb{E}[\| \int_0^t \int_H (S_{t-s} f(s, u, Z_s) - S_{t-s}^n f(s, u, Z_s^n)) q(ds, du)\|^2].$$

We shall analyze separately the three terms on the right-hand side of inequality (5.3.4). As for the first term, we remark that

$$\lim_{n \to \infty} \|S_t^n \xi - S_t \xi\| = 0.$$

By Sect. 5.1 (Yosida approximation) we get that the convergence is uniform in $[0, T]$.

Let us consider the second term on the right-hand side of (5.3.4). We have:

$$\mathbb{E}[\| \int_0^t (S_{t-s} a(s, Z_s) - S_{t-s}^n a(s, Z_s^n)) ds\|^2]$$
$$\leq 2T \int_0^t \mathbb{E}[\|S_{t-s} a(s, Z_s) - S_{t-s}^n a(s, Z_s)\|^2] ds$$
$$+ 2T \int_0^t \mathbb{E}[\|S_{t-s}^n a(s, Z_s) - S_{t-s}^n a(s, Z_s^n)\|^2] ds \tag{5.3.5}$$

$$\lim_{n \to \infty} \|S_{t-s} a(s, Z_s(\omega)) - S_{t-s}^n a(s, Z_s(\omega))\| = 0 \quad P - a.s. \tag{5.3.6}$$

and

$$\|S_{t-s} a(s, Z_s(\omega)) - S_{t-s}^n a(s, Z_s(\omega))\|^2 \leq C_T \|a(s, Z_s(\omega))\|^2$$
$$\leq C_T K (\|Z_s(\omega)\|^2 + 1). \tag{5.3.7}$$

This is a consequence of uniform convergence and condition (D). By the Lebesgue dominated convergence theorem it follows that the first term on the r.h.s. of (5.3.5) converges to zero.

Let us consider the second term on the r.h.s. of (5.3.5). We observe that from uniform convergence and the Lipschitz condition (C) it follows that

$$T\|S_{t-s}^n a(s, Z_s(\omega)) - S_{t-s}^n a(s, Z_s^n(\omega))\|^2 \le C_T L \|Z_s(\omega) - Z_s^n(\omega)\|^2$$

so that

$$2T \int_0^t \mathbb{E}[\|S_{t-s}^n a(s, Z_s) - S_{t-s}^n a(s, Z_s^n)\|^2\, ds \le 2C_T L \int_0^t \mathbb{E}[\|Z_s - Z_s^n\|^2]\, ds.$$

It follows that for all $\epsilon > 0$ there is an $n_0 \in \mathbb{N}$ such that for all $n \ge n_0$

$$\mathbb{E}[\|\int_0^t (S_{t-s} a(s, Z_s) - S_{t-s}^n a(s, Z_s^n)) ds\|^2] \le \epsilon + 2C_T L \int_0^t \mathbb{E}[\|Z_s - Z_s^n\|^2]\, ds.$$

Let us consider the third term in (5.3.4). By similar arguments as above, it can be proved that

$$\mathbb{E}[\|\int_0^t \int_H (S_{t-s} f(s, u, Z_s) - S_{t-s}^n f(s, u, Z_s^n)) q(ds, du)\|^2]$$
$$\le \epsilon + 2C_T L \int_0^t \mathbb{E}[\|Z_s - Z_s^n\|^2]\, ds.$$

It follows that

$$\mathbb{E}[\|Z_t - Z_t^n\|^2] \le 2^3 \|S_t^n \xi - S_t \xi\|^2 + 2^4 \epsilon 2^4 C_T L \int_0^t \mathbb{E}[\|Z_s - Z_s^n\|^2]\, ds.$$

Using Gronwall's Lemma we get

$$\mathbb{E}[\|Z_t - Z_t^n\|^2] \le (2^3 \|S_t^n \xi - S_t \xi\|^2 + 2^4 \epsilon) \exp(2^4 T L C_T)$$

so that (5.3.4) gives the result. □

5.4 The Markov Property of the Solution of SPDEs

Let $B_b(H)$ denote the set of bounded real valued functions on H. We first prove that the Markov property holds for the semigroup associated to the mild solutions of (5.3.1):

Let $0 < v < T$ and $\xi \in H$. Let $(Z(t, v, \xi))_{t \in [v,T]}$ denote the solution of the following integral equation

$$Z_t = S_{t-v}\xi + \int_v^t S_{t-s}a(s, Z_s)ds + \int_v^t \int_H S_{t-s}f(s, u, Z_s)q(ds, du) \quad (5.4.1)$$

(in the sense of Theorem 5.3.1). Let $\mathcal{F}_t^Z$ denote the σ-algebra generated by $Z(\tau, v, \xi)$, with $\tau \leq t, \tau \geq v$. Let $v \leq s \leq t \leq T$ and $P_{s,t}$ be the linear operator on $B_b(H)$, defined by

$$(P_{s,t})(\phi)(x) = \mathbb{E}[\phi(Z(t, s; x))] \quad \text{for} \quad \phi \in B_b(H) \quad x \in H. \quad (5.4.2)$$

Then the Markov property holds, i.e.

Theorem 5.4.1 *Let $0 \leq v \leq s \leq t \leq T$. Then*

$$\mathbb{E}[\phi(Z(t, v; \xi))/\mathcal{F}_s^Z] = (P_{s,t})(\phi)(Z(s, v; \xi)) \text{ for any } \phi \in B_b(H).$$

Proof As $\mathcal{F}_s^Z \subset \mathcal{F}_s$, it is sufficient to prove that

$$\mathbb{E}[\phi(Z(t, v; \xi))/\mathcal{F}_s] = (P_{s,t})(\phi)(Z(s, v; \xi)). \quad (5.4.3)$$

From the uniqueness of the solution we get

$$Z(t, v; \xi)(\omega) = Z(t, s; Z(s, v; \xi)(\omega))(\omega) \quad \mathbb{P} - a.s. \quad (5.4.4)$$

Let

$$\eta(\omega) := Z(s, v; \xi)(\omega). \quad (5.4.5)$$

Then from (5.4.4) it follows that (5.4.3) can be written as

$$\mathbb{E}[\phi(Z(t, s; \eta))/\mathcal{F}_s] = (P_{s,t})(\phi)(Z(s, v; \eta)). \quad (5.4.6)$$

It is enough to show that (5.4.6) holds for every $\phi \in C_b(H)$, with $C_b(H)$ denoting the set of continuous real-valued bounded functions on H. We first assume that ϕ is linear and bounded.

Moreover, let us first consider the case where

$$\eta(\omega) = x \in H \quad \mathbb{P} - a.s.$$

As x is constant and because of the independent increment property of the cPrm, $Z(t, s; \eta(\omega))$ is independent of $\mathcal{F}_s$. In fact $\mathcal{F}_s$ is the σ-algebra generated by the pure jump Lévy process with compensator $ds\beta(dx)$. See Sect. 2.4 and Sect. 3.3, or [3, Sect. 2].

$$\mathbb{E}[\phi(Z(t, s; \eta))/\mathcal{F}_s] = \mathbb{E}[\phi(Z(t, s, x))] = P_{s,t}(\phi(x))$$

so that (5.4.6) holds for this particular case.

Now we prove (5.4.6) for the case where

$$\eta(\omega) := \sum_1^n a_j \mathbf{1}_{A_j}(Z(s, v; \xi)) \tag{5.4.7}$$

with $\{A_j, j = 1, \ldots, n\}$ a partition of H and $a_1, \ldots, a_n \in H$. In this case

$$Z(t, s; \eta(\omega))(\omega) = \sum_1^n Z(t, s; a_j)\mathbf{1}_{A_j}(Z(s, v; \xi)) \quad \mathbb{P} - a.s.,$$

$$\phi(Z(t, s; \eta(\omega))(\omega)) = \sum_1^n \phi(Z(t, s; a_j))\mathbf{1}_{A_j}(Z(s, v; \xi)) \quad \mathbb{P} - a.s.,$$

and

$$\mathbb{E}[\phi(Z(t, s; \eta)/\mathcal{F}_s] = \mathbb{E}[\sum_1^n \phi(Z(t, s; a_j))\mathbf{1}_{A_j}(Z(s, v; \xi))/\mathcal{F}_s]$$

$$= \sum_1^n P_{s,t}(\phi)(a_j)\mathbf{1}_{A_j}(Z(s, v, \xi)) = P_{s,t}(\phi)(\eta), \tag{5.4.8}$$

where in (5.4.8) we used that $\phi(Z(t, s; a_j))$ are independent of $\mathcal{F}_s$ and $\mathbf{1}_{A_j}(Z(s, v; \xi))$ are $\mathcal{F}_s$-measurable.

Now we prove (5.4.6) for the case where $\eta(\omega)$ is given according to (5.4.5). (From the proof it follows in particular that the r.h.s. of (5.4.3) is $\mathcal{F}_s^Z$-measurable.) There is a sequence of simple functions $\eta_n(\omega)$ of the form (5.4.7) such that, if for a given natural number M we denote $\eta_n^M := \eta_n \wedge M$, then

$$\lim_{M \to \infty} \lim_{n \to \infty} \mathbb{E}[\|\eta_n^M - \eta\|^2] = 0. \tag{5.4.9}$$

Similar to the proof of Corollary 5.3.2 it follows that

$$\lim_{M \to \infty} \lim_{n \to \infty} \mathbb{E}[\|Z(t, s; \eta_n^M) - Z(t, s; \eta)\|^2] = 0.$$

There is a subsequence (by abuse of notation we denote it in the same way as the original sequence), for which

$$\lim_{M \to \infty} \lim_{n \to \infty} Z(t, s; \eta_n^M)(\omega) = Z(t, s; \eta)(\omega) \quad \mathbb{P} - a.s.$$

As ϕ is continuous and bounded, it follows from (5.4.8) that

$$\mathbb{E}[\phi(Z(t, s; \eta)/\mathcal{F}_s)] = \lim_{M \to \infty} \lim_{n \to \infty} \mathbb{E}[\phi(Z(t, s, \eta_n^M)/\mathcal{F}_s)]$$

$$= \lim_{M \to \infty} \lim_{n \to \infty} P_{s,t}(\phi)(\eta_n^M) = P_{s,t}(\phi)(\eta).$$

Given $\phi \in C_b(H)$ there exists a sequence of linear bounded functions ϕ_n converging, up to a set of Borel measure zero, to ϕ (see e.g. [103], Chap. V.5). It follows that $\phi_n(Z(t,s;\eta)) \to \phi(Z(t,s;\eta))$ $\mathbb{P}$-a.s., when $n \to \infty$. ϕ_n can be chosen to be uniformly bounded, so that

$$\lim_{n\to\infty} \mathbb{E}[\phi_n(Z(t,s;\eta)/\mathcal{F}_s)] = \mathbb{E}[\phi(Z(t,s;\eta)/\mathcal{F}_s)]. \qquad \square$$

Theorem 5.4.2 *Let $T > 0$, $f(s,u,z)) = f(u,z)$, $a(s,z) = a(z)$ and $x \in H$, then $(Z(t,0;x)(\omega))_{t\in[0,T]}$ is a homogenous Markov process.*

Proof It is sufficient to prove that

$$P_{s,t} = P_{0,t-s} \quad \text{for all} \quad 0 \le s \le t \le T \qquad (5.4.10)$$

together with the Markov property in Theorem 5.4.1 implies that the Chapman–Kolmogorov equation holds for the transition probabilities associated to $P_{s,t}, 0 \le s \le t \le T$ and $(Z(t,0;x)(\omega))_{t\in[0,T]}$ is a Markov process.

Let us remark that the compensated Lévy random measure $q(ds,du)(\omega)$ is translation invariant in time, i.e. if $t > 0$ and $\tilde{q}(ds,du)(\omega)$ denotes the unique σ-finite measure on $\mathcal{B}(\mathbb{R}_+ \times H)$ which extends the pre-measure $\tilde{q}(ds,du)(\omega)$ on $S(\mathbb{R}_+) \times \mathcal{B}(H)$, such that $\tilde{q}((s,\tau], \Lambda) := q((s+t,\tau+t], \Lambda)$, for $(s,\tau] \times \Lambda \in S(\mathbb{R}_+) \times \mathcal{B}(H)$, then $\tilde{q}(B)$ and $q(B)$ are equally distributed for all $B \in \mathcal{B}(\mathbb{R}_+ \times H)$.

It follows that

$$Z(t+h,t;x)$$
$$= S_h x + \int_t^{t+h} S_{t+h-s} a(Z(s,t;x))ds + \int_t^{t+h}\int_H S_{t+h-s} f(Z(s,t;x))q(ds,du)$$
$$= S_h x + \int_0^h S_{h-s} a(Z(t+s,t;x))ds + \int_0^h\int_H S_{h-s} f(u,Z(t+s,t;x))\tilde{q}(ds,du)$$
$$= S_h x + \int_0^h S_{h-s} a(Z(t+s,t;x))ds + \int_0^h\int_H S_{h-s} f(u,Z(t+s,t;x))q(ds,du).$$

By uniqueness (Theorem 5.3.1) it follows that $Z(t+h,t;x)(\omega)$ and $Z(h,0;x)(\omega)$ have the same distribution, completing the proof. $\qquad \square$

5.5 Existence of Solutions for Random Coefficients

Let $L_2^T := L_2^T([0,T] \times \Omega, (\mathcal{F}_t)_{t\in[0,T]})$ be the space of processes $(Z_t(\omega))_{t\in[0,T]}$ which are jointly measurable and

(i) Z_t is $\mathcal{F}_t$-measurable,
(ii) $\int_0^T \mathbb{E}[\|Z_s\|^2]ds < \infty$.

Definition 5.5.1 We say that two processes $Z_t^i(\omega) \in L_2^T$, $i = 1, 2$, are $dt \otimes P$-equivalent if they coincide for all $(t, \omega) \in \Gamma$, with $\Gamma \in \mathcal{B}([0, T]) \otimes \mathcal{F}_T$, and $dt \otimes P(\Gamma^c) = 0$. We denote by $\mathcal{L}_2^T$ the set of $dt \otimes P$-equivalence classes.

Remark 5.5.2 $\mathcal{L}_2^T$, with norm

$$\|Z_t\|_{\mathcal{L}_2^T} := \left(\int_0^T \mathbb{E}[\|Z_s\|^2] ds \right)^{1/2},$$

is a Hilbert space.

In this section we assume that the coefficients are random and adapted to the filtration and prove the existence of a solution in $\mathcal{L}_2^T$. We assume here the growth and Lipschitz conditions of the coefficients independent of ω, but depending on the points in H. We assume in fact that we are given

$$a : \mathbb{R}_+ \times H \times \Omega \to H,$$

$$f : \mathbb{R}_+ \times H \times H \times \Omega \to H,$$

such that
(A') $f(t, u, z, \omega)$ is jointly measurable such that for all $t \in [0, T]$, $u \in E$ and fixed $z \in H$, $f(t, u, z, \cdot)$ is $\mathcal{F}_t$-adapted,
(B') $a(t, z, \omega)$ is jointly measurable such that for all $t \in [0, T]$, and fixed $z \in H$, $a(t, z, \cdot)$ is $\mathcal{F}_t$-adapted, and for fixed $T > 0$
(C') there is a constant $L > 0$ such that

$$T\|a(t, z, \omega) - a(t, z', \omega)\|^2 + \int_H \|f(t, u, z, \omega) - f(t, u, z', \omega)\|^2 \beta(du) \le L\|z - z'\|^2$$

for all $t \in [0, T]$, $z, z' \in H$, and $\mathbb{P} - a.e. \ \omega \in \Omega$,

(D') there is a constant $K > 0$ such that

$$T\|a(t, z, \omega)\|^2 + \int_H \|f(t, u, z, \omega)\|^2 \beta(du) \le K(\|z\|^2 + 1)$$
for all $t \in ([0, T]$, $z \in H$, and $\mathbb{P} - a.e. \ \omega \in \Omega$.

Theorem 5.5.3 *Let $0 < T < \infty$ and suppose that (A'), (B'), (C'), (D') are satisfied. Let $x \in H$. Then there is a unique process $(Z_t)_{0 \le t \le T} \in \mathcal{L}_2^T$ which satisfies*

$$Z_t(\omega) = S_t x + \int_0^t S_{t-s} a(s, Z_s(\omega), \omega) ds$$

$$+ \int_0^t \int_H S_{t-s} f(s, u, Z_s(\omega), \omega) q(ds du) \quad \forall t \in [0, T]. \quad (5.5.1)$$

As a consequence of Theorem 5.5.3 we have:

Corollary 5.5.4 *Let $0 < T < \infty$ and suppose that (A'), (B'), (C'), (D') are satisfied. Then there is up to stochastic equivalence a unique process $(Z_t)_{0 \leq t \leq T} \in L_2^T$ which satisfies (5.5.1).*

Remark 5.5.5 As a consequence of Lemma 5.1.9 we have that $(Z_t)_{0 \leq t \leq T}$ is càdlàg.

Before proving Theorem 5.5.3 we prove some properties of the following function

$$K_t(x, \xi)(\omega) := S_t x + \int_0^t S_{t-s} a(s, \xi_s(\omega), \omega) ds$$

$$+ \int_0^t \int_H S_{t-s} f(s, u, \xi_s(\omega), \omega) q(ds, du)$$

with $x \in H$ and $\xi := (\xi_s)_{s \in [0,T]} \in L_2^T$.

Lemma 5.5.6 *For any $T > 0$ there is a constant C_T^1 such that*

$$\int_0^T \mathbb{E}[\|K_t(x, \xi) - K_t(x, \eta)\|^2] dt \leq C_T^1 \int_0^T \mathbb{E}[\|\xi_t - \eta_t\|^2] dt.$$

Proof

$$\int_0^T \mathbb{E}[\|K_t(x, \xi) - K_t(x, \xi)\|^2] dt$$

$$\leq 2e^{2\alpha T} T \int_0^T \mathbb{E}\left[\left\| \int_0^t (a(s, \xi_s) - a(s, \eta_s)) ds \right\|^2 \right] dt$$

$$+ 2e^{2\alpha T} \int_0^T \int_0^t \int_H \mathbb{E}[\|(f(s, u, \xi_s) - f(s, u, \eta_s)\|^2 ds \beta(du)] dt$$

$$\leq 2LT e^{2\alpha T} \int_0^T \mathbb{E}[\|\xi_s - \eta_s\|^2] dt < \infty,$$

where we applied the bounds on S_t. This proves the lemma.   □

Let

$$K(x, \xi) : H \times L_2^T \rightarrow L_2^T \tag{5.5.2}$$

be such that its projection at time $t \in [0, T]$ is given by $K_t(x, \xi)$.

Lemma 5.5.7 *There exists a constant α_T, depending on T, such that $\alpha_T \in (0, 1)$ and*

$$\|K(x, \xi)(\omega) - K(x, \eta)(\omega)\|_{L_2^T} \leq \alpha_T \|\xi - \eta\|_{L_2^T}. \tag{5.5.3}$$

Proof Let $S\xi := K_t(x, \xi)$. We shall prove that $\mathbf{S}^n$ is a contraction operator on $\mathcal{L}_2^T$, for sufficiently large values of $n \in \mathbb{N}$. By Lemma 5.5.6 it follows by induction that

$$\int_0^T \mathbb{E}[\|\mathbf{S}^n \xi_t - \mathbf{S}^n \eta_t\|^2] dt \leq C_T^{1^n} \int_0^T dt \int_0^T ds_1 \int_0^T ds_2, \ldots, \int_0^T \mathbb{E}[\|\xi_{s_n} - \eta_{s_n}\|^2] ds_n$$

$$\leq C_T^{1^n} \frac{T^n}{n!} \int_0^T \mathbb{E}[\|\xi_s - \eta_s\|^2] ds.$$

From this we get that, for sufficiently large values of $n \in \mathbb{N}$, the operator $\mathbf{S}^n$ is a contraction operator on $\mathcal{L}_2^T$ and therefore has a unique fixed point. Suppose that $\mathbf{S}^{n_0}$ is a contraction operator on $\mathcal{L}_2^T$. We get

$$\int_0^T dt \mathbb{E}[\|\mathbf{S}\xi_t - \mathbf{S}\eta_t\|^2] = \int_0^T dt \mathbb{E}[\|\mathbf{S}^{kn_0+1}\xi_t - \mathbf{S}^{kn_0+1}\eta_t\|^2]$$

$$\leq \frac{C_T^{1^{kn_0}} T^{kn_0}}{kn_0!} \int_0^T dt \mathbb{E}[\|\mathbf{S}\xi_t - \mathbf{S}\eta_t\|^2]$$

$$\leq \frac{C_T^{1^{kn_0+1}} T^{kn_0}}{kn_0 + 1!} \int_0^T dt \mathbb{E}[\|\xi_t - \eta_t\|^2] \to 0 \quad \text{as} \quad k \to \infty.$$

$\square$

Proof of Theorem 5.5.3 From (5.5.3) it follows that $K(x, \xi)$ is a contraction on $\mathcal{L}_2^T$ for every fixed $x \in H$. We get by the contraction principle that there exists a $\phi \in C(H, \mathcal{L}_2^T)$ such that

$$K(x, \phi(x)) = \phi(x)$$

for every fixed $x \in H$. $\phi(x) := (Z_t^x(\omega))_{t \in [0,T]}$ is the solution of (5.5.1). $\square$

5.6 Continuous Dependence on Initial Data, Drift and Noise Coefficients

Let $T > 0$. Let us assume that (A), (B), (C), (D) or (A′), (B′), (C′), (D′) are satisfied for $f_0(t, u, z, \omega) := f(t, u, z, \omega)$ and $a_0(t, z, \omega) := a(t, z, \omega)$. Moreover, we assume that this also holds for $f_n(t, u, z, \omega)$ and $a_n(t, z, \omega)$, for any $n \in \mathbb{N}$. Let $(Z_t)_{t \in [0,T]}$ be a solution of (5.5.1) (in the sense of the previous theorems, depending on the hypothesis). We denote by $(Z_t^n(\omega))_{[0,T]}$ the unique solution of

$$Z_t^n(\omega) = S_t Z_0^n(\omega) + \int_0^t S_{t-s} a_n(s, Z_s^n(\omega), \omega) ds$$

$$+ \int_0^t \int_H S_{t-s} f_n(s, u, Z_s^n(\omega), \omega) q(ds, du)$$

(in the sense of the previous theorems). We prove the following result:

Theorem 5.6.1 *Assume that there is a constant $K > 0$ such that for all $n \in \mathbb{N}_0$, $t \in [0, T]$ and $z \in H$*

$$\|a_n(t, z, \omega)\|^2 + \int_H \|f_n(t, u, z, \omega)\|^2 \beta(du) \leq K(\|z\|^2 + 1) \quad \mathbb{P}-a.s. \quad (5.6.1)$$

Assume that there is a constant L such that for all $n \in \mathbb{N}_0$, $t \in [0, T]$ and $z, z' \in H$:

$$T\|a_n(t, z, \omega) - a_n(t, z', \omega)\|^2 + \int_H \|f_n(t, u, z, \omega) - f_n(t, u, z', \omega)\|^2 \beta(du)$$
$$\leq L\|z - z'\|^2 \quad \mathbb{P}-a.s. \quad (5.6.2)$$

Moreover, assume that

$$\sup_{n \in \mathbb{N}_0} \mathbb{E}[\|Z_0^n)\|^2] < \infty, \quad (5.6.3)$$

$$\lim_{n \to \infty} \mathbb{E}[\|Z_0^n - Z_0\|^2] = 0 \quad (5.6.4)$$

(where $Z_0^0(\omega) := Z_0(\omega)$) and assume that for every $t \in [0, T]$ and fixed $z \in H$

$$\lim_{n \to \infty} \{T\|a_n(t, z, \omega) - a(t, z, \omega)\|^2 + \int_H \|f_n(t, u, z, \omega) - f(t, u, z, \omega)\|^2 \beta(du)\}$$
$$= 0 \quad \mathbb{P}-a.s. \quad (5.6.5)$$

Then

$$\lim_{n \to \infty} \sup_{t \in [0,T]} \mathbb{E}[\|Z_t^n - Z_t\|^2] = 0.$$

Proof Let $t \leq T$, then:

$$\mathbb{E}[\|Z_t^n - Z_t\|^2] \leq 2^5 e^{2\alpha T} \{\mathbb{E}[\|Z_0^n - Z_0\|^2] + 2L \int_0^t \mathbb{E}[\|Z_t^n - Z_t\|^2]ds$$
$$+ 2T \int_0^t \mathbb{E}[\|a_n(s, Z_s) - a(s, Z_s)\|^2]ds\}$$
$$+ 2^5 e^{2\alpha T} \{2 \int_0^t \int_H \mathbb{E}[\|f_n(s, u, Z_s) - f(s, u, Z_s)\|^2] \beta(du)ds\},$$

where the latter inequality is proved by using a bound on $\|S_t\|$ and inequality (5.6.2).

Let

$$\gamma_t^n := T \int_0^t \mathbb{E}[\|a_n(s, Z_s) - a(s, Z_s)\|^2]ds,$$

$$\delta_t^n := \int_0^t \int_H \mathbb{E}[\|f_n(s, u, Z_s) - f(s, u, Z_s)\|^2]\beta(du)ds.$$

As

$$\lim_{n \to \infty} \|a_n(s, Z_s, \omega) - a(s, Z_s, \omega)\|^2$$
$$+ \int_H \|f_n(s, u, Z_s, \omega) - f(s, u, Z_s, \omega)\|^2\beta(du) = 0, \quad \mathbb{P} - a.s.$$

and (5.6.1) implies

$$\|a_n(t, Z_s(\omega), \omega)\|^2 + \int_H \|f_n(t, u, Z_s(\omega), \omega)\|^2\beta(du)$$
$$\leq K(\|Z_s(\omega)\|^2 + 1) \quad \mathbb{P} - a.s.,$$

it follows that

$$\lim_{n \to \infty} \sup_{t \in [0,T]} \delta_t^n + \lim_{n \to \infty} \sup_{t \in [0,T]} \gamma_t^n = 0.$$

The conclusion then follows by using Gronwall's inequality. □

5.7 Differential Dependence of the Solutions on the Initial Data

In this section we continue to assume, as before, that the coefficients a and f satisfy the conditions (A), (B), (C) and (D) and we shall prove the differential dependence of the solution of (5.3.1) with respect to the initial data. Let

$$K_t(x, \xi) := S_t x + \int_0^t S_{t-s} a(s, \xi_s)ds + \int_0^t \int_H S_{t-s} f(s, u, \xi_s)q(ds, du)$$

with $x \in H$ and $\xi := (\xi_s)_{s \in [0,T]} \in \mathcal{L}_2^T$.

Lemma 5.7.1 *For any $T > 0$ there is a constant C_T^1, resp. C_T^2, such that*

$$\int_0^T \mathbb{E}[\|K_t(x, \xi) - K_t(x, \eta)\|^2]dt \leq C_T^1 \int_0^T \mathbb{E}[\|\xi_t - \eta_t\|^2]dt, \quad (5.7.1)$$

$$\int_0^T \mathbb{E}[\| K_t(x, \xi) - K_t(y, \xi) \|^2] dt \le C_T^2 \| x - y \|^2. \qquad (5.7.2)$$

Proof Note that (5.7.1) is a special case of Lemma 5.5.6. The proof of (5.7.2) is similar to that of Lemma 5.2.2. □

Let

$$K(x, \xi) : H \times \mathcal{L}_2^T \to \mathcal{L}_2^T$$

be such that its projection at time $t \in [0, T]$ is given by $K_t(x, \xi)$.

Remark 5.7.2 From Theorem 5.3.1 we know that there is a unique solution $(Z_t^x (\omega))_{t \in [0,T]}$ of (5.3.1). Hence, from Theorem 5.3.1 we know that for every fixed $x \in H$

$$K(x, Z_t^x(\omega)) = Z_t^x(\omega) \quad \mathbb{P} - a.s. \qquad (5.7.3)$$

We shall now prove some facts about the map K.

Theorem 5.7.3 *Let $\xi \in \mathcal{L}_2^T$ be fixed. The map*

$$K(\cdot, \xi) : H \to \mathcal{L}_2^T$$

is Fréchét differentiable and its derivative $\frac{\partial K}{\partial x}$ along the direction $h \in H$ is such that

$$\frac{\partial K_t(x, \xi)}{\partial x}(h) = S_t h.$$

The proof of Theorem 5.7.3 is easy and follows from the Frechét differentiability of S_t.

Remark 5.7.4 It follows in particular that $\frac{\partial K}{\partial x}$ is in $\mathcal{L}(H; \mathcal{L}_2^T)$.

Let us denote by $\frac{\partial}{\partial z}$ the Fréchét derivative in H. Starting from here we assume that the coefficients a and f in the SPDE also satisfy the following conditions
(E) $\frac{\partial}{\partial z} f(t, u, z)$ exists for all $t \in (0, T]$ and fixed $u \in H$,
(F) $\frac{\partial}{\partial z} a(t, z)$ exists for all $t \in (0, T]$.

Moreover we assume that

$$\||\frac{\partial}{\partial z} a(s, z)|\|^2 + \int_H \||\frac{\partial}{\partial z} f(s, z, u)|\|^2 \beta(du) < \infty \quad \text{uniformly in} \quad z \in H,$$

$$\text{and} \quad s \in [0, T], \quad (5.7.4)$$

where $\|| \cdot \||$ denotes the operator norm of the Fréchét derivative in H.

Theorem 5.7.5 *Let $x \in H$ be fixed.*

$$K(x, \cdot) : \mathcal{L}_2^T \to \mathcal{L}_2^T \tag{5.7.5}$$

is Gateaux differentiable and its derivative $\frac{\partial K}{\partial \xi}$ along the direction $\xi \in \mathcal{L}_2^T$ satisfies

$$\frac{\partial K_t(x, \xi)}{\partial \xi}(\eta_t) = \int_0^t S_{t-s} \frac{\partial}{\partial z} a(s, \xi_s)(\eta_s) ds$$

$$+ \int_0^t \int_H S_{t-s} \frac{\partial}{\partial z} f(s, u, \xi_s)(\eta_s) q(ds, du)$$

(with the notation $\frac{\partial}{\partial z} a(s, \xi_s(\omega))$ (resp. $\frac{\partial}{\partial z} f(s, u, \xi_s(\omega)))$ for $\frac{\partial}{\partial z} a(s, z)$ (resp. $\frac{\partial}{\partial z} f(s, u, z)$), at $z = \xi_s(\omega)$).

Proof For any fixed $x \in H$, and $\xi, \eta \in \mathcal{L}_2^T$ we consider the map $r \to K(x, \xi + r\eta)$ from $\mathbb{R}$ to $\mathcal{L}_2^T$. We have

$$K_t(x, \xi + r\eta) = S_t x + \int_0^t S_{t-s} a(s, \xi_s + r\eta_s) ds$$

$$+ \int_0^t \int_H S_{t-s} f(s, u, \xi_s + r\eta_s) q(ds, du).$$

It follows that

$$\frac{1}{r}(K_t(x, \xi + r\eta) - K(x, \xi)) = \int_0^t S_{t-s} \frac{(a(s, \xi_s + r\eta_s) - a(s, \xi_s))}{r} ds$$

$$+ \int_0^t \int_H S_{t-s} \frac{(f(s, u, \xi_s + r\eta_s) - f(s, u, \xi_s))}{r} q(ds, du).$$

Let us fix $z \in H$ and define for any $r \neq 0$:

$$a_r(t, z, y) := \frac{a(t, z + ry) - a(t, z)}{r}$$

$$f_r(t, u, z, y) := \frac{f(t, u, z + ry) - f(t, u, z)}{r}$$

where $t \in [0, T]$, $y \in H$. $a_r(t, y, \xi_s(\omega))$ and $f_r(t, u, y, \xi_s(\omega))$ satisfy the conditions (5.6.1) and (5.6.2) with r instead of n (and y instead of z). Moreover, $\frac{\partial}{\partial z} a(s, \xi_s(\omega))y$ and $\frac{\partial}{\partial z} f(s, u, \xi_s(\omega))y$ satisfy the same conditions, by condition (5.7.4).

Analogous to (5.6.5), we have (also using the Lipschitz conditions) that

$$\lim_{r\to 0} \{T\|a_r(t, y, \xi_t(\omega)) - \frac{\partial}{\partial z}a(t, \xi_t(\omega))y\|^2 + \int_{E\setminus\{0\}} \|f_r(t, u, y, \xi_t(\omega))$$
$$- \frac{\partial}{\partial z}f(t, u, \xi_t(\omega))y\|^2\beta(du)\} = 0 \quad \mathbb{P}-a.s.$$

Defining similarly as before

$$\gamma_t^r := T\int_0^t \mathbb{E}[\|a_r(s, \eta_s, \xi_s,) - \frac{\partial}{\partial z}a(s, \xi_s)\eta_s\|^2]ds$$

$$\delta_t^r := \int_0^t \int_H \mathbb{E}[\|f_r(s, u, \eta_s, \xi_s) - \frac{\partial}{\partial z}f(s, u, \xi_s)\eta_s\|^2]\beta(du)ds,$$

and operating in a similar way as in the proof of Theorem 5.6.1, we obtain the desired result.   □

We also assume
(G) $\frac{\partial}{\partial z}a(s, z)$ is continuous in z ds-a.s.
(H) $\frac{\partial}{\partial z}f(s, u, z)$ is continuous ds-a.s. in the norm $\|\cdot\|_{\mathcal{L}^2(d\beta)}$ of $\mathcal{L}^2(d\beta)$.

Theorem 5.7.6 *For any fixed $\eta \in \mathcal{L}_2^T$ the function*

$$\frac{\delta}{\delta\xi}K(x, \xi)\eta : H \times \mathcal{L}_2^T \to \mathcal{L}_2^T \tag{5.7.6}$$

is continuous.

Proof of Theorem 5.7.6 Let (x^n, ξ^n) converge to (x, ξ) in $H \times \mathcal{L}_2^T$. For any $n \in \mathbb{N}$ we have that

$$\frac{\partial}{\partial\xi}K(x^n, \xi^n)\eta_t - \frac{\partial}{\partial\xi}K(x, \xi)\eta_t = \int_0^t S_{t-s}(\frac{\partial}{\partial z}a(s, \xi_s^n)\eta_s - \frac{\partial}{\partial z}a(s, \xi_s)\eta_s)$$
$$+ \int_0^t \int_H S_{t-s}(\frac{\partial}{\partial z}f(s, u, \xi_s^n)\eta_s$$
$$- \frac{\partial}{\partial z}f(s, u, \xi_s)\eta_s)q(ds, dx).$$

From $\|S_t\| \le e^{\alpha t}$ it follows that

$$\int_0^T \mathbb{E}[\|\frac{\partial}{\partial\xi}K(x^n, \xi^n)\eta_t - \frac{\partial}{\partial\xi}K(x, \xi)\eta_t\|^2]dt$$

$$\le 2Te^{2\alpha T}\int_0^T \mathbb{E}[\|\frac{\partial}{\partial z}a(s, \xi_s^n)\eta_s - \frac{\partial}{\partial z}a(s, \xi_s)\eta_s\|^2]ds$$

$$+ 2T e^{2\alpha T} \int_0^T \int_H \mathbb{E}[\| \frac{\partial}{\partial z} f(s, u, \xi_s^n)\eta_s - \frac{\partial}{\partial z} f(s, u, \xi_s)\eta_s \|^2] ds \, \beta(du).$$

$$(5.7.7)$$

$\xi^n \to \xi$ in $\mathcal{L}_2^T$ as $n \to \infty$ implies that there is a subsequence $\{n_k\}_{k \in \mathbb{N}}$ such that $\xi_s^{n_k} \to \xi_s \, ds \otimes d\mathbb{P}$ -a.s. in $[0, T] \times \Omega$, as $k \to \infty$. Hence we have

$$\| \frac{\partial}{\partial z} a(s, \xi_s^{n_k}(\omega))\eta_s - \frac{\partial}{\partial z} a(s, \xi_s(\omega))\eta_s \| \to 0 \quad ds \otimes d\mathbb{P} - a.e.$$

$$\text{in } [0, T] \times \Omega \text{ as } k \to \infty$$

and

$$\int_H \| \frac{\partial}{\partial z} f(s, u, \xi_s^n(\omega))\eta_s - \frac{\partial}{\partial z} f(s, u, \xi_s(\omega))\eta_s \|^2 \beta(du) \to 0$$

$$a.e. \ ds \otimes d\mathbb{P} \text{ in } [0, T] \times \Omega. \tag{5.7.8}$$

We get by the Lebesgue dominated convergence theorem that $\frac{\partial}{\partial \xi} K(x, \xi)\eta$ is continuous. □

Corollary 5.7.7 *Let us assume that all the hypotheses of Theorem 5.7.6 hold. Let $(Z_t^x)_{t \in [0,T]}$ denote the solution of (5.3.1) with initial condition*

$$Z_0(\omega) = x \quad \mathbb{P} - a.s.$$

Then $(\frac{\partial}{\partial x} Z_t^x)_{t \in [0,T]}$ is a solution of

$$\frac{\partial}{\partial x} Z_t^x = \int_0^t (S_{t-s} \frac{\partial}{\partial z} a(s, Z_s^x) \frac{\partial}{\partial x} Z_s^x) \, ds$$

$$+ \int_0^t \int_H (S_{t-s} \frac{\partial}{\partial z} f(s, u, Z_s^x) \frac{\partial}{\partial x} Z_s^x) \, q(ds, dx). \tag{5.7.9}$$

Proof The statement of Corollary 5.7.7 is a consequence of Theorems 5.7.3–5.7.6, Remark 5.7.4 and Proposition C.0.3 in Appendix C of [15] (see also Appendix C of [19], where the Gaussian case is considered). □

5.8 Remarks and Related Literature

In this chapter, we have studied Hilbert space valued SPDEs. A special case of SPDEs in Banach spaces with certain restrictions on the partial differential operator was considered in [38].

Our presentation is based on [2]. The technique is a generalization of that used in [35] (see also [34]) and is generalized from [36].

The material on Gateaux differentiability with respect to the initial value was generalized in [72].

This work has found applications to financial models in [32].

For our work in Chap. 7 on stability theory we provide the Yosida approximations for mild solutions. As the approximating solutions are strong solutions, we can apply Itô's formula for these. The general case of non-anticipating coefficients is of interest in view of the applications presented in [24].

We refer the reader to [33] where Sz.-Nagy's dilation theorem is used to study uniqueness by relating mild solutions to strong solutions. However, in this form, one does not know how to study the asymptotic behaviour of the equation in Sect. 6.1 to obtain the result on invariant measure in [71], which is done in Chap. 7.

Chapter 6
Applications

In this chapter we show how the results of Chap. 5 can be used to solve some problems arising in finance. In addition, we provide motivation for the study of Chap. 5 since the Zakai equation in filtering problems has the form of the SPDEs studied there.

6.1 The HJMM Equation from Interest Rate Theory

In this section we describe the HJMM model for term structure interest rates. We follow [32] in our presentation in the next section. We start by explaining some fundamental ideas from finance mathematics. After deriving the HJMM equation, we consider the existence and uniqueness of this equation using our results in Chap. 5 on mild solutions of SPDEs. Under additional assumptions on the space of forward curves and drift, we obtain the strong solution result of [17]. In order to demonstrate the strength of our result, we present an example for which the assumptions in [17] are not satisfied. In the next section we introduce the basic financial problem. We do not give full details of the work as it is already described in [32]. However, we give sufficient details to motivate our model.

6.1.1 Introduction to the HJMM Equation

A zero coupon bond with maturity T is a financial asset which pays the holder one unit of cash at time T. Its price for $t \leq T$ can be written as

$$p(t, T) = \exp\left(-\int_t^T f(t, u)du\right)$$

where $f(t, T)$ is the forward rate at time T ($\geq t$).

© Springer International Publishing Switzerland 2015
V. Mandrekar and B. Rüdiger, *Stochastic Integration in Banach Spaces*,
Probability Theory and Stochastic Modelling 73, DOI 10.1007/978-3-319-12853-5_6

The classical continuous framework for the evolution of forward rates goes back to Heath et al. [40]. They assume that under a risk-neutral measure for every time T the forward rates $f(t, T)$ follow an Itô process of the form

$$df(t, T) = \sum_{i=1}^{n} \sigma_i(t, T) \int_t^T \sigma(t, s)ds \qquad (6.1.1)$$

$$+ \sum_{i=1}^{n} \sigma_i(t, T)dW_t^i, \quad t \in [0, T] \qquad (6.1.2)$$

where $W = (W^1, \ldots, W^n)$ is a standard Brownian motion in $\mathbb{R}^n$. This gives that the discounted zero coupon bond price processes

$$\exp\left(-\int_0^t f(s, s)ds\right) p(t, T) \quad t \in [0, T]$$

are local martingales for all maturities T. This guarantees absence of arbitrage in the bond market model.

Empirical studies have revealed that models based on a Brownian motion (noise) only provide a poor fit to observed market data [31, Chap. 5]. Some authors [12, 26], and others have proposed to replace Brownian motion W in (6.1.1) by a more general process (with jumps). If X is a Lévy process, this leads to

$$df(t, T) = \alpha_{HJM}(t, T)dt + \sum_{i=1}^{n} \sigma_i(t, T)dX_t^i, \quad t \in [0, T]. \qquad (6.1.3)$$

Here the drift term is replaced in (6.1.1) by an appropriate term determined by $\sigma(t, T)$ and a generating function of X (as explained later).

From a financial modeling point of view, one considers σ and α_{HJM} to be a function of the prevailing forward curve $T \to f(t-, T, \omega) = \lim_{s \uparrow t} f(s, T, \omega)$. This leads to $f(t, T)$ being a solution of the stochastic equation for $t \in [0, T]$

$$df(t, T) = \alpha_{HJM}(t, T, f(t, \cdot))dt + \sum_{i=1}^{n} \sigma_i(t, T, f(t, \cdot))dX_t^i, \quad t \in [0, T] \quad (6.1.4)$$

where $f(0, T) = h_0(T)$ is some initial forward curve.

Let us switch to an alternative parametrization to show that Eq. (6.1.4) is an infinitesimal stochastic PDE. Let us introduce the Musiela parametrization [78]

$$r_t(x) = f(t, t + x), \quad x \geq 0.$$

Then the above equation in integrated form becomes

$$r_t = S_t h_0(x) + \int_0^t S_{t-s} \alpha_{HJM}(s, s+x, r_s) dt + \sum_{i=1}^n \int_0^t S_{t-s} \sigma_i(s, s+x, r_s) dX_s^i$$

$$(6.1.5)$$

where $S_t h = h(t + \cdot)$ for $t \in \mathbb{R}_+$, that is (see Chap. 5), r_t is a mild solution of the equation

$$dr_t = \frac{d}{dx} r_t(x) + \alpha_{HJM}(t, r_t) dt + \sum_{i=1}^n \sigma_i(t, r_t) dX_t^i \qquad (6.1.6)$$

in an appropriate Hilbert space H of forward curves, where $\frac{d}{dx}$ is the generator of the strongly continuous semigroup S_t (shift).

Here we use (with a slight abuse of notation):

$\alpha_{HJM}(t, r.)$ for $\alpha_{HJM}(t, t + \cdot, r.)$ and $\sigma(t, r.)$ for $\sigma_{HJM}(t, t + \cdot, r.)$.

The advantage of using the representation (6.1.6) instead of (6.1.4) is that instead of dealing with infinitely many SDEs, one for every maturity time T, we can deal with only one (infinite-dimensional) SPDE.

Before we specify H in various cases, we end this section by motivating the HJM drift condition.

Throughout H denotes a separable Hilbert space of forward curves and σ_i : $\mathbb{R}_+ \times H \rightarrow H$ $(i = 1, 2, \ldots, n)$ are volatilities. In order that the term structure model (6.1.6) is free of arbitrage, we have to show that all discounted bond prices are local martingales. In order to achieve this, we assume that these are compact intervals $[a_1, b_1], [a_2, b_2], \ldots, [a_n, b_n]$ having zero as inner point, such that the Lévy measures $\nu_1, \nu_2, \ldots, \nu_n$ of $X^1, X^2, \ldots, X^n$ respectively, satisfy for $i = 1, 2, \ldots, n$

$$\int_{|x|>1} e^{zx} \nu_i(dx) < \infty \quad \text{for} \quad z \in [a_i, b_i]. \qquad (6.1.7)$$

From (6.1.7) we see that

$$\psi_i(z) = \ln E[e^{zX^i}] \quad i = 1, \ldots, n$$

exists on $[a_i, b_i]$ and is in C^∞ [95]. Moreover, the Lévy processes X^i possess moments of arbitrary order. Let $[c_i, d_i] \subset (a_i, b_i)$ be compact intervals having zero as an inner point. For any continuous function $h : \mathbb{R}_+ \rightarrow \mathbb{R}$ define $Th : \mathbb{R}_+ \rightarrow \mathbb{R}$ by

$$Th(x) = \int_0^x h(\eta) d\eta.$$

For $i = 1, 2, \ldots, n$, let

$$A_H^{\psi_i} = \{h \in H; -Th(\mathbb{R}_+) \subset [c_i, d_i]\}.$$

If $\sigma_i(\mathbb{R}_+ \times H) \subset A_H^{\psi_i}, i = 1, 2, \ldots, n$, the HJM drift is

$$\alpha_{HJM}(t, r)(x) = \sum_{i=1}^{n} \frac{d}{dx} \psi_i \left(-\int_0^x \sigma_i(t, r)(\eta) d\eta \right) \tag{6.1.8}$$

$$= -\sigma_i(t, r)(x) \psi_i' \left(-\int_0^x \sigma_i(t, r)(\eta) d\eta \right) \tag{6.1.9}$$

which is well defined for all x. The HJM drift condition above implies that P is a local martingale measure [31, Sect. 2.1].

Remark 6.1.1 We need to ensure that $\alpha_{HJM}(t, r) \in H$ for all $(t, r) \in \mathbb{R} \times H$ and that the Lipschitz property of σ_i ($i = 1, 2, \ldots, n$) implies the Lipschitz property of α_{HJM}. This requires us to choose the space of forward curves carefully. In order that (6.1.5) implies (6.1.6) we also need the map $h \mapsto h(x)$ from H to $\mathbb{R}$ to be continuous.

6.1.2 The Space of Forward Curves and Mild Solution to the HJJM Equation

In this section we introduce the space of forward curves following [31]. We shall present the existence and uniqueness result for the case $n = 1$. In view of our condition on the existence of all moments of the Lévy measure, we can incorporate the Poisson part of the Lévy decomposition of the Lévy process occurring in Eq. (6.1.6) into the drift part, observing that the Lipschitz condition on σ implies the Lipschitz condition needed in Theorem 5.2.3.

Now we consider the existence of mild and weak solutions to (6.1.6). We first define the spaces H_w of forward curves, which were introduced in [31, Chap. 5].

Let $w : \mathbb{R}_+ \to [1, \infty)$ be a non-decreasing C^1-function such that $w^{-\frac{1}{3}} \in L^1(\mathbb{R}_+)$.

Example 6.1.2 $w(x) = e^{\alpha x}$, for $\alpha > 0$.

Example 6.1.3 $w(x) = (1 + x)^\alpha$, for $\alpha > 3$.

Let H_w be the linear space of all absolutely continuous functions $h : \mathbb{R}_+ \to \mathbb{R}$ satisfying

$$\int_{\mathbb{R}_+} |h'(x)|^2 w(x) dx < \infty,$$

where h' denotes the weak derivative of h. We define the inner product

$$(g, h)_w := g(0)h(0) + \int_{\mathbb{R}_+} g'(x)h'(x)w(x)dx$$

and denote the corresponding norm by $\||\cdot\||_w$. Since, for large time, forward curves flatten to the maturity x, the choice of H_w is reasonable from an economic point of view.

Proposition 6.1.4 *The space* $(H_w, (\cdot, \cdot)_w)$ *is a separable Hilbert space. Each* $h \in H_w$ *is continuous, bounded and the limit* $h(\infty) := \lim_{x \to \infty} h(x)$ *exists. Moreover, for each* $x \in \mathbb{R}_+$, *the point evaluation* $h \mapsto h(x) : H_w \to \mathbb{R}$ *is a continuous linear functional.*

Proof All of these statements can be found in the proof of [31, Theorem 5.1.1]. $\square$

The fact that each point evaluation is a continuous linear functional ensures that forward curves (r_t) solving (6.1.6) satisfy the variation of constants formula (6.1.5). Defining the constants $C_1, \ldots, C_4 > 0$ as

$$C_1 := \|w^{-1}\|_{L^1(\mathbb{R}_+)}^{\frac{1}{2}}, \quad C_2 := 1 + C_1, \quad C_3 := \|w^{-\frac{1}{3}}\|_{L^1(\mathbb{R}_+)}^{2},$$

$$C_4 := \|w^{-\frac{1}{3}}\|_{L^1(\mathbb{R}_+)}^{\frac{7}{2}},$$

we have for all $h \in H_w$ the estimates

$$\|h'\|_{L^1(\mathbb{R}_+)} \le C_1 \||h\||_w, \tag{6.1.10}$$

$$\|h\|_{L^\infty(\mathbb{R}_+)} \le C_2 \||h\||_w, \tag{6.1.11}$$

$$\|h - h(\infty)\|_{L^1(\mathbb{R}_+)} \le C_3 \||h\||_w, \tag{6.1.12}$$

$$\|(h - h(\infty))^4 w\|_{L^1(\mathbb{R}_+)} \le C_4 \||h\||_w^4, \tag{6.1.13}$$

which also follows by inspecting the proof of [31, Theorem 5.1.1].

Since in order to apply Theorem 5.2.3 we require that the shift semigroup $(S_t)_{t \ge 0}$ defined by $S_t h = h(t + \cdot)$ for $t \in \mathbb{R}_+$ is pseudo-contractive in a closed subspace of H_w, we use a technique which is due to Tehranchi [98], namely we change to the inner product

$$\langle g, h \rangle_w := g(\infty)h(\infty) + \int_{\mathbb{R}_+} g'(x)h'(x)w(x)dx$$

and denote the corresponding norm by $\|\cdot\|_w$. The estimates (6.1.10)–(6.1.13) are also valid with the norm $\|\cdot\|_w$ for all $h \in H_w$ (the proof is exactly as for the original norm $\||\cdot\||_w$). Therefore we conclude, by using (6.1.11),

$$\frac{1}{(1+C_2^2)^{\frac{1}{2}}}\|h\|_w \leq \||h\||_w \leq (1+C_2^2)^{\frac{1}{2}}\|h\|_w, \quad h \in H_w$$

showing that $\|\cdot\|_w$ and $\||\cdot\||_w$ are equivalent norms on H_w. From now on, we shall work with the norm $\|\cdot\|_w$.

Proposition 6.1.5 (S_t) *is a C_0-semigroup in H_w with generator* $\frac{d}{dx} : \mathcal{D}(\frac{d}{dx}) \subset H_w \to H_w$, $\frac{d}{dx}h = h'$, *and domain*

$$\mathcal{D}(\tfrac{d}{dx}) = \{h \in H_w \,|\, h' \in H_w\}.$$

The subspace $H_w^0 := \{h \in H_w \,|\, h(\infty) = 0\}$ *is a closed subspace of H_w and (S_t) is contractive in H_w^0 with respect to the norm $\|\cdot\|_w$.*

Proof Except for the last statement, we refer to the proof of [31, Theorem 5.1.1]. By the monotonicity of w we have

$$\|S_t h\|_w^2 = \int_{\mathbb{R}_+} |h'(x+t)|^2 w(x)dx \leq \|h\|_w^2$$

for all $t \in \mathbb{R}_+$ and $h \in H_w^0$, showing that (S_t) is contractive in H_w^0. □

We define for any $h = (h_1, \ldots, h_n) \in \times_{i=1}^n A_{H_w^0}^{\Psi_i}$

$$\Sigma h(x) := -\sum_{i=1}^n h_i(x)\Psi_i'\left(-\int_0^x h_i(\eta)d\eta\right), \quad x \in \mathbb{R}_+. \tag{6.1.14}$$

Proposition 6.1.6 *There is a constant $C_5 > 0$ such that for all $g, h \in \times_{i=1}^n A_{H_w^0}^{\Psi_i}$ we have*

$$\|\Sigma g - \Sigma h\|_w \leq C_5 \sum_{i=1}^n \left(1 + \|h_i\|_w + \|g_i\|_w + \|g_i\|_w^2\right)\|g_i - h_i\|_w. \tag{6.1.15}$$

Furthermore, for each $h \in \times_{i=1}^n A_{H_w^0}^{\Psi_i}$ we have $\Sigma h \in H_w^0$, and the map $\Sigma : \times_{i=1}^n A_{H_w^0}^{\Psi_i} \to H_w^0$ is continuous.

Proof We define

$$K_i := \sup_{x\in[c_i,d_i]} |\Psi_i'(x)|, \quad L_i := \sup_{x\in[c_i,d_i]} |\Psi_i''(x)| \quad \text{and} \quad M_i := \sup_{x\in[c_i,d_i]} |\Psi_i'''(x)|$$

for $i = 1, \ldots, n$. By the boundedness of the derivatives Ψ_i' on $[c_i, d_i]$, the definition (6.1.14) of Σ yields that for each $h \in \times_{i=1}^n A_{H_w^0}^{\Psi_i}$ the limit $\Sigma h(\infty) := \lim_{x\to\infty} \Sigma h(x)$ exists and

$$\Sigma h(\infty) = 0, \quad h \in \times_{i=1}^{n} A_{H_w^0}^{\Psi_i}. \tag{6.1.16}$$

By using (6.1.16) and the universal inequality

$$|x_1 + \cdots + x_k|^2 \leq k\left(|x_1|^2 + \cdots + |x_k|^2\right), \quad k \in \mathbb{N}$$

we get for arbitrary $g, h \in \times_{i=1}^{n} A_{H_w^0}^{\Psi_i}$ the estimate

$$\|\Sigma g - \Sigma h\|_w^2 = \int_{\mathbb{R}_+} \left| \sum_{i=1}^{n} h_i'(x)\Psi_i'\left(-\int_0^x h_i(\eta)d\eta\right) \right.$$

$$- \sum_{i=1}^{n} g_i'(x)\Psi_i'\left(-\int_0^x g_i(\eta)d\eta\right) + \sum_{i=1}^{n} g_i(x)^2\Psi_i''\left(-\int_0^x g_i(\eta)d\eta\right)$$

$$\left. - \sum_{i=1}^{n} h_i(x)^2\Psi_i''\left(-\int_0^x h_i(\eta)d\eta\right) \right|^2 w(x)dx \leq 4n(I_1 + I_2 + I_3 + I_4),$$

where we have put

$$I_1 := \sum_{i=1}^{n} \int_{\mathbb{R}_+} |h_i'(x)|^2 \left| \Psi_i'\left(-\int_0^x h_i(\eta)d\eta\right) - \Psi_i'\left(-\int_0^x g_i(\eta)d\eta\right) \right|^2 w(x)dx,$$

$$I_2 := \sum_{i=1}^{n} \int_{\mathbb{R}_+} \Psi_i'\left(-\int_0^x g_i(\eta)d\eta\right)^2 |h_i'(x) - g_i'(x)|^2 w(x)dx,$$

$$I_3 := \sum_{i=1}^{n} \int_{\mathbb{R}_+} g_i(x)^4 \left[\Psi_i''\left(-\int_0^x g_i(\eta)d\eta\right) - \Psi_i''\left(-\int_0^x h_i(\eta)d\eta\right) \right]^2 w(x)dx,$$

$$I_4 := \sum_{i=1}^{n} \int_{\mathbb{R}_+} \Psi_i''\left(-\int_0^x h_i(\eta)d\eta\right)^2 (g_i(x)^2 - h_i(x)^2)^2 w(x)dx.$$

Using (6.1.12) yields

$$I_1 \leq \sum_{i=1}^{n} L_i^2 \|h_i\|_w^2 \|g_i - h_i\|_{L^1(\mathbb{R}_+)}^2 \leq C_3^2 \sum_{i=1}^{n} L_i^2 \|h_i\|_w^2 \|g_i - h_i\|_w^2,$$

and I_2 is estimated as

$$I_2 \leq \sum_{i=1}^{n} K_i^2 \|g_i - h_i\|_w^2.$$

Taking into account (6.1.12) and (6.1.13), we get

$$I_3 \leq \sum_{i=1}^n M_i^2 \|g_i^4 w\|_{L^1(\mathbb{R}_+)} \|g_i - h_i\|_{L^1(\mathbb{R}_+)}^2 \leq C_3^2 C_4 \sum_{i=1}^n M_i^2 \|g_i\|_w^4 \|g_i - h_i\|_w^2,$$

and by using Hölder's inequality and (6.1.13), we obtain

$$I_4 \leq \sum_{i=1}^n L_i^2 \int_{\mathbb{R}_+} (g_i(x) + h_i(x))^2 w(x)^{\frac{1}{2}} (g_i(x) - h_i(x))^2 w(x)^{\frac{1}{2}} dx$$

$$\leq \sum_{i=1}^n L_i^2 \|(g_i + h_i)^4 w\|_{L^1(\mathbb{R}_+)}^{\frac{1}{2}} \|(g_i - h_i)^4 w\|_{L^1(\mathbb{R}_+)}^{\frac{1}{2}}$$

$$\leq 2C_4 \sum_{i=1}^n L_i^2 (\|g_i\|_w^2 + \|h_i\|_w^2) \|g_i - h_i\|_w^2,$$

which gives us the desired estimate (6.1.15). For all $h \in \times_{i=1}^n A_{H_w^0}^{\Psi_i}$ we have $\Sigma h \in H_w^0$ by (6.1.15), (6.1.16), and the map $\Sigma : \times_{i=1}^n A_{H_w^0}^{\Psi_i} \to H_w^0$ is locally Lipschitz continuous by (6.1.15). □

By Proposition 6.1.6 we can, for given volatilities $\sigma_i : \mathbb{R}_+ \times H_w \to H_w^0$ satisfying $\sigma_i(\mathbb{R}_+ \times H_w) \subset A_{H_w^0}^{\Psi_i}$ for $i = 1, \ldots, n$, define the drift term α_{HJM} according to the HJM drift condition (6.1.8) by

$$\alpha_{HJM} := \Sigma \circ \sigma : \mathbb{R}_+ \times H_w \to H_w^0, \tag{6.1.17}$$

where $\sigma = (\sigma_1, \ldots, \sigma_n)$.

Now, we are ready to establish the existence of Lévy term structure models on the space H_w of forward curves.

Theorem 6.1.7 Let $\sigma_i : \mathbb{R}_+ \times H_w \to H_w^0$ be continuous and satisfy $\sigma_i(\mathbb{R}_+ \times H_w) \subset A_{H_w^0}^{\Psi_i}$ for $i = 1, \ldots, n$. Assume there are $M, L \geq 0$ such that for all $i = 1, \ldots, n$ and $t \in \mathbb{R}_+$ we have

$$\|\sigma_i(t, h)\|_w \leq M, \quad h \in H_w$$
$$\|\sigma_i(t, h_1) - \sigma_i(t, h_2)\|_w \leq L\|h_1 - h_2\|_w, \quad h_1, h_2 \in H_w.$$

Then, for each $h_0 \in H_w$, there exists a unique mild and a unique weak adapted càdlàg solution $(r_t)_{t\geq 0}$ to (6.1.6) with $r_0 = h_0$ satisfying

$$\mathbb{E}\left[\sup_{t\in[0,T]} \|r_t\|_w^2 \right] < \infty \quad \text{for all } T > 0. \tag{6.1.18}$$

Proof By Proposition 6.1.6, α_{HJM} maps into H_w^0, see (6.1.17). Since $\sigma = (\sigma_1, \ldots, \sigma_n) : \mathbb{R}_+ \times H_w \to \times_{i=1}^n A_{H_w^0}^{\Psi_i}$ is continuous by assumption and $\Sigma : \times_{i=1}^n A_{H_w^0}^{\Psi_i} \to H_w^0$ is continuous by Proposition 6.1.6, it follows that $\alpha_{\text{HJM}} = \Sigma \circ \sigma$ is continuous. Moreover, by estimate (6.1.15), we obtain for all $t \in \mathbb{R}_+$ and $h_1, h_2 \in H_w$ the estimate

$$\|\alpha_{\text{HJM}}(t, h_1) - \alpha_{\text{HJM}}(t, h_2)\|_w \leq C_5(1+M)^2 \sum_{i=1}^n \|\sigma_i(t, h_1) - \sigma_i(t, h_2)\|_w$$

$$\leq C_5(1+M)^2 nL \|h_1 - h_2\|_w.$$

Taking into account Proposition 6.1.5, applying Theorem 5.2.3 completes the proof. $\qquad\square$

As an immediate consequence, we get the existence of Lévy term structure models with constant direction volatilities.

Corollary 6.1.8 *Let* $\sigma_i : \mathbb{R}_+ \times H_w \to H_w^0$ *be defined by* $\sigma_i(t, r) = \sigma_i(r) = \varphi_i(r)\lambda_i$, *where* $\lambda_i \in A_{H_w^0}^{\Psi_i}$ *and* $\varphi_i : H_w \to [0, 1]$ *for* $i = 1, \ldots, n$. *Assume there is an* $L \geq 0$ *such that for all* $i = 1, \ldots, n$ *we have*

$$|\varphi_i(h_1) - \varphi_i(h_2)| \leq L \|h_1 - h_2\|_w, \quad h_1, h_2 \in H_w.$$

Then, for each $h_0 \in H_w$, *there exists a unique mild and a unique weak adapted càdlàg solution* $(r_t)_{t \geq 0}$ *to* (6.1.6) *with* $r_0 = h_0$ *satisfying* (6.1.18).

Proof For all $h_1, h_2 \in H_w$ and all $i = 1, \ldots, n$ we have

$$\|\sigma_i(h_1) - \sigma_i(h_2)\|_w \leq L \|\lambda_i\|_w \|h_1 - h_2\|_w.$$

Observing that $\|\sigma_i(h)\|_w \leq \|\lambda_i\|_w$ for all $h \in H_w$ and $i = 1, \ldots, n$, the proof is a straightforward consequence of Theorem 6.1.7. $\qquad\square$

The only assumption on the driving Lévy processes $X^1, \ldots, X^n$, needed to apply the previous results, is the exponential moments condition (6.1.7). It is clearly satisfied for Brownian motions and Poisson processes.

There are also several purely discontinuous Lévy processes fulfilling (6.1.7), for instance generalized hyperbolic processes, which have been introduced by Barndorff-Nielsen [8], and their subclasses, namely the normal inverse Gaussian and hyperbolic processes. These processes have been applied to finance by Eberlein and co-authors in a series of papers, e.g. in [25].

Other purely discontinuous Lévy processes satisfying (6.1.7) are the generalized tempered stable processes, see [17, Sect. 4.5], which include Variance Gamma processes [61], and bilateral Gamma processes [57].

Consequently, Theorem 6.1.7 applies to term structure models driven by any of the above types of Lévy processes.

6.1.3 Forward Curve Evolutions as Strong Solutions of Infinite-Dimensional Stochastic Differential Equations

In this section we choose the forward curves as in [13] and show that in this case $\frac{d}{dx}$ is a bounded operator on the space. This allows us to obtain strong solutions to Eq. (6.1.6) using Theorem 4.2.2, as the space of forward curves is a Hilbert space (clearly of M-type 2). We remark that under assumption (6.1.7) the Poisson integral part of Eq. (6.1.6) (using Lévy decomposition) can be replaced by compensated Poisson, adding the compensating term to the drift.

We fix real numbers $\beta > 1$ and $\gamma > 0$. Let $H_{\beta,\gamma}$ be the linear space of all $h \in C^\infty(\mathbb{R}_+, \mathbb{R})$ satisfying

$$\sum_{n=0}^{\infty} \left(\frac{1}{\beta}\right)^n \int_0^\infty \left(\frac{d^n h(x)}{dx^n}\right) e^{-\gamma x} dx < \infty.$$

We define the inner product

$$\langle g, h \rangle_{\beta,\gamma} := \sum_{n=0}^{\infty} \left(\frac{1}{\beta}\right)^n \int_0^\infty \left(\frac{d^n g(x)}{dx^n}\right)\left(\frac{d^n h(x)}{dx^n}\right) e^{-\gamma x} dx$$

and denote the corresponding norm by $\| \cdot \|_{\beta,\gamma}$. From [13, Proposition 4.2] we obtain the following Propositions 6.1.9 and 6.1.10.

Proposition 6.1.9 *The space $(H_{\beta,\gamma}, \langle \cdot, \cdot \rangle_{\beta,\gamma})$ is a separable Hilbert space and for each $x \in \mathbb{R}_+$, the point evaluation $h \mapsto h(x) : H_{\beta,\gamma} \to \mathbb{R}$ is a continuous linear functional.*

The fact that each point evaluation is a continuous linear functional ensures that forward curves (r_t) solving (6.1.6) satisfy the variation of constants formula (6.1.5).

Proposition 6.1.10 *We have $\frac{d}{dx} \in \mathcal{L}(H_{\beta,\gamma})$, i.e. $\frac{d}{dx}$ is a bounded linear operator on $H_{\beta,\gamma}$.*

Theorem 6.1.11 *Let $\sigma_i : \mathbb{R}_+ \times H_{\beta,\gamma} \to H_{\beta,\gamma}$ be continuous and satisfy $\sigma_i(\mathbb{R}_+ \times H_{\beta,\gamma}) \subset A_{H_{\beta,\gamma}}^{\Psi_i}$ for $i = 1, \ldots, n$. Assume that $\alpha_{HJM}(t, r) \in H_{\beta,\gamma}$ for all $(t, r) \in \mathbb{R}_+ \times H_{\beta,\gamma}$. Furthermore, assume that $\alpha_{HJM}(t, r) : \mathbb{R}_+ \times H_{\beta,\gamma} \to H_{\beta,\gamma}$ is continuous and that there is a constant $L \geq 0$ such that for all $t \in \mathbb{R}_+$ and $h_1, h_2 \in H_{\beta,\gamma}$ we have*

$$\|\alpha_{HJM}(t, h_1) - \alpha_{HJM}(t, h_2)\|_{\beta,\gamma} \leq L \|h_1 - h_2\|_{\beta,\gamma},$$
$$\|\sigma_i(t, h_1) - \sigma_i(t, h_2)\|_{\beta,\gamma} \leq L \|h_1 - h_2\|_{\beta,\gamma}, \quad i = 1, \ldots, n.$$

Then, for each $h_0 \in H_{\beta,\gamma}$, there exists a unique strong adapted càdlàg solution $(r_t)_{t \geq 0}$ to (6.1.6) with $r_0 = h_0$ satisfying

$$\mathbb{E}\left[\sup_{t\in[0,T]} \|r_t\|_{\beta,\gamma}^2\right] < \infty \quad \text{for all } T > 0. \tag{6.1.19}$$

Proof Taking into account Proposition 6.1.10, the result is a consequence of Corollary 4.2.2. ∎

Unfortunately, Theorem 6.1.11 has some shortcomings, namely it is demanded that the drift term α_{HJM} according to the HJM drift condition maps into the space $H_{\beta,\gamma}$. The following simple counterexample shows that this condition may be violated.

Example 6.1.12 Let $\sigma = -1$ and X be a compound Poisson process with intensity $\lambda = 1$ and jump size distribution $N(0,1)$. Notice that the compound Poisson process satisfies the exponential moments condition (6.1.7) for all $z \in \mathbb{R}$, because its Lévy measure is given by

$$F(dx) = \frac{1}{\sqrt{2\pi}} e^{-\frac{x^2}{2}} dx.$$

But we have $\alpha_{\mathrm{HJM}} \notin H_{\beta,\gamma}$, because

$$\int_0^\infty \alpha_{\mathrm{HJM}}(x)^2 e^{-\gamma x} dx = \int_0^\infty \left(\frac{d}{dx}\Psi(x)\right)^2 e^{-\gamma x} dx$$

$$= \int_0^\infty \left(\frac{d}{dx}\left(e^{\frac{x^2}{2}} - 1\right)\right)^2 e^{-\gamma x} dx = \int_0^\infty x^2 e^{x^2 - \gamma x} dx = \infty.$$

The phenomena that the drift α_{HJM} may be located outside the space of forward curves $H_{\beta,\gamma}$ has to do with the fact that the space $H_{\beta,\gamma}$ is a very small space in a sense, in particular, every function must necessarily be real-analytic (see [13, Proposition 4.2]).

The small size of this space arises from the requirement that $\frac{d}{dx}$ should be a bounded operator, because we are dealing with the existence of *strong* solutions. When dealing with *mild* solutions, problems of this kind will not occur.

Nevertheless, for certain types of term structure models, we can apply Theorem 6.1.11. For this purpose, we proceed with a lemma. For a given real-analytic function $h : \mathbb{R}_+ \to \mathbb{R}$ it is, in general, difficult to decide whether h belongs to $H_{\beta,\gamma}$ or not. For the following functions this information can be provided.

Lemma 6.1.13 *Every polynomial p belongs to $H_{\beta,\gamma}$, and for $\delta \in \mathbb{R}$ satisfying $\delta^2 < \beta$ and $\delta < \frac{\gamma}{2}$, the function $h(x) = e^{\delta x}$ belongs to $H_{\beta,\gamma}$.*

Proof The first statement is clear. For $h(x) = e^{\delta x}$ we obtain

$$\sum_{n=0}^{\infty} \left(\frac{1}{\beta}\right)^n \int_0^{\infty} \left(\frac{d^n h(x)}{dx^n}\right)^2 e^{-\gamma x} dx = \sum_{n=0}^{\infty} \left(\frac{1}{\beta}\right)^n \int_0^{\infty} \left(\delta^n e^{\delta x}\right)^2 e^{-\gamma x} dx$$

$$= \sum_{n=0}^{\infty} \left(\frac{\delta^2}{\beta}\right)^n \int_0^{\infty} e^{-(\gamma-2\delta)x} dx$$

$$= \frac{1}{1 - \frac{\delta^2}{\beta}} \cdot \frac{1}{\gamma - 2\delta}$$

$$= \frac{\beta}{(\beta - \delta^2)(\gamma - 2\delta)},$$

whence $h \in H_{\beta,\gamma}$. $\qquad\square$

Let $n = 3$, that is, we have three independent driving processes. We denote by X^1 and X^2 two standard Wiener processes, and X^3 is a Poisson process with intensity $\lambda > 0$. We specify the volatilities as

$$\sigma_1(r)(x) = \varphi_1(r)p(x), \quad \sigma_2(r)(x) = \varphi_2(r)e^{\delta x} \text{ and } \sigma_3(r)(x) = -\eta, \qquad (6.1.20)$$

where p is a polynomial, $\delta, \eta \in \mathbb{R}$ satisfy $4\delta^2 < \beta$, $\delta < \frac{\gamma}{4}$ and $\eta^2 < \beta$, $\eta < \frac{\gamma}{2}$, and where $\varphi_i : H_{\gamma,\beta} \to \mathbb{R}$ for $i = 1, 2$. Note that $\sigma_i(H_{\beta,\gamma}) \subset H_{\beta,\gamma}$ for $i = 1, 2, 3$ by Lemma 6.1.13. The drift according to the HJM drift condition (6.1.8) is given by

$$\alpha_{\text{HJM}}(r)(x) = \frac{d}{dx}\left[\frac{1}{2}\varphi_1(r)^2 q(x)^2 + \frac{1}{2}\varphi_2(r)^2 \left(\frac{e^{\delta x} - 1}{\delta}\right)^2 + \lambda\left(e^{\eta x} - 1\right)\right],$$

where $q(x) = \int_0^x p(\eta)d\eta$ is again a polynomial. From Lemma 6.1.13 and Proposition 6.1.10 we infer $\alpha_{\text{HJM}}(H_{\beta,\gamma}) \subset H_{\beta,\gamma}$.

Proposition 6.1.14 *Assume there is a constant $L \geq 0$ such that for all $h_1, h_2 \in H_{\beta,\gamma}$ we have*

$$|\varphi_i(h_1) - \varphi_i(h_2)| \leq L\|h_1 - h_1\|_{\beta,\gamma}, \quad i = 1, 2,$$

$$|\varphi_i(h_1)^2 - \varphi_i(h_2)^2| \leq L\|h_1 - h_1\|_{\beta,\gamma}, \quad i = 1, 2.$$

Then, for each $h_0 \in H_{\beta,\gamma}$, there exists a unique strong adapted càdlàg solution $(r_t)_{t \geq 0}$ to (6.1.6) with $r_0 = h_0$ satisfying (6.1.19).

Proof We have for all $h_1, h_2 \in H_{\beta,\gamma}$

$$\|\sigma_1(h_1) - \sigma_1(h_2)\| \leq L\|p\|_{\beta,\gamma}\|h_1 - h_2\|_{\beta,\gamma},$$

$$\|\sigma_2(h_1) - \sigma_2(h_2)\| \leq L\|e^{\delta \bullet}\|_{\beta,\gamma}\|h_1 - h_2\|_{\beta,\gamma}.$$

Using Proposition 6.1.10, we obtain for all $h_1, h_2 \in H_{\beta,\gamma}$

$$\|\alpha_{HJM}(h_1) - \alpha_{HJM}(h_2)\|_{\beta,\gamma} \leq \frac{L}{2} \|A\|_{\mathcal{L}(H_{\beta,\gamma})} \left(\|q^2\|_{\beta,\gamma} \right.$$
$$\left. + \|\tfrac{1}{\delta^2}(e^{\delta \bullet} - 1)^2\|_{\beta,\gamma} \right) \|h_1 - h_2\|_{\beta,\gamma}.$$

Applying Theorem 6.1.11 completes the proof.   □

In order to generalize Proposition 6.1.14, by allowing that η in (6.1.20) may depend on the present state of the forward curve, instead of being constant, we prepare two auxiliary results.

Lemma 6.1.15 *Let $\gamma > 0$ and $g, h \in C^1(\mathbb{R}_+; \mathbb{R})$. Assume there are $c > 0$, $\varepsilon \in (-\infty, \gamma)$ and $x_0 \in \mathbb{R}_+$ such that*

$$|g(x)h(x)| \leq c e^{\varepsilon x} \quad \text{for all } x \geq x_0.$$

Then we have

$$\int_0^\infty g(x)h(x)e^{-\gamma x}dx = \frac{1}{\gamma}\left[g(0)h(0) + \int_0^\infty g'(x)h(x)e^{-\gamma x}dx \right.$$
$$\left. + \int_0^\infty g(x)h'(x)e^{-\gamma x}dx \right].$$

Proof Performing partial integration with three factors, we obtain

$$\left[g(x)h(x)e^{-\gamma x} \right]_0^\infty = \int_0^\infty g'(x)h(x)e^{-\gamma x}dx + \int_0^\infty g(x)h'(x)e^{-\gamma x}dx$$
$$- \gamma \int_0^\infty g(x)h(x)e^{-\gamma x}dx.$$

By hypothesis, we have $\lim_{x\to\infty} g(x)h(x)e^{-\gamma x} = 0$, and so the stated formula follows.   □

Lemma 6.1.16 *Let $\gamma > 0$ and $h \in C^2(\mathbb{R}_+; \mathbb{R})$ be such that $h, h', h'' \geq 0$. Assume there are $c > 0$, $\varepsilon \in (-\infty, \frac{\gamma}{2})$ and $x_0 \in \mathbb{R}_+$ such that*

$$|h(x)| \leq c e^{\varepsilon x} \text{ and } |h'(x)| \leq c e^{\varepsilon x} \quad \text{for all } x \geq x_0.$$

Then we have

$$\int_0^\infty h'(x)^2 e^{-\gamma x}dx \leq \frac{\gamma^2}{2} \int_0^\infty h(x)^2 e^{-\gamma x}dx.$$

Proof Using Lemma 6.1.15 twice, we obtain

$$\int_0^\infty h(x)^2 e^{-\gamma x} dx = \frac{2}{\gamma} \int_0^\infty h(x)h'(x)e^{-\gamma x}dx + \frac{1}{\gamma}h(0)^2$$

$$= \frac{2}{\gamma^2}\left[\int_0^\infty h'(x)^2 e^{-\gamma x}dx + \int_0^\infty h(x)h''(x)e^{-\gamma x}dx\right]$$

$$+ \frac{1}{\gamma}\left[h(0)^2 + \frac{2}{\gamma}h(0)h'(0)\right].$$

Since $h, h', h'' \geq 0$ by hypothesis, the stated inequality follows. $\square$

Now we generalize Proposition 6.1.14 by assuming that, instead of being constant, $\eta : H_{\beta,\gamma} \to \mathbb{R}$ in (6.1.20) is allowed to depend on the current state of the forward curve. The rest of our present framework is exactly as in Proposition 6.1.14.

Proposition 6.1.17 *Assume that, in addition to the hypothesis of Proposition 6.1.14, we have $\gamma \leq \sqrt{2}$, $\eta(H_{\beta,\gamma}) \subset [0,\frac{\gamma}{2}) \cap [0,\sqrt{\beta})$ and*

$$|\eta(h_1) - \eta(h_2)| \leq L\|h_1 - h_2\|_{\beta,\gamma}$$

for all $h_1, h_2 \in H_{\beta,\gamma}$. Then, for each $h_0 \in H_{\beta,\gamma}$, there exists a unique strong adapted càdlàg solution $(r_t)_{t\geq 0}$ to (6.1.6) with $r_0 = h_0$ satisfying (6.1.19).

Proof It suffices to show that $\Gamma : H_{\beta,\gamma} \to H_{\beta,\gamma}$ defined as $\Gamma(r)(x) := e^{\eta(r)x}$ is Lipschitz continuous. So let $h_1, h_2 \in H_{\beta,\gamma}$ be arbitrary. Without loss of generality we assume that $\eta(h_2) \leq \eta(h_1)$. Observe that all derivatives of $\Gamma(h_1) - \Gamma(h_2)$ are non-negative. So we obtain by applying Lemma 6.1.16 (notice that $\gamma \leq \sqrt{2}$ by hypothesis), and the Lipschitz property $|e^x - e^y| \leq e^x|x - y|$ for $y \leq x$, that

$$\|\Gamma(h_1) - \Gamma(h_2)\|_{\beta,\gamma}^2 = \sum_{n=0}^\infty \left(\frac{1}{\beta}\right)^n \int_0^\infty \left(\eta(h_1)^n e^{\eta(h_1)x} - \eta(h_2)^n e^{\eta(h_2)x}\right)^2 e^{-\gamma x}dx$$

$$\leq \frac{\beta}{\beta-1}\int_0^\infty \left(e^{\eta(h_1)x} - e^{\eta(h_2)x}\right)^2 e^{-\gamma x}dx$$

$$\leq \frac{\beta}{\beta-1}\int_0^\infty \left(e^{\eta(h_1)x}(\eta(h_1) - \eta(h_2))x\right)^2 e^{-\gamma x}dx$$

$$\leq \frac{\beta}{\beta-1}\left(\int_0^\infty \left(xe^{\eta(h_1)x}\right)^2 e^{-\gamma x}dx\right)L^2\|h_1 - h_2\|_{\beta,\gamma}^2.$$

The integral is finite, because we have $\eta(h_1) \in [0,\frac{\gamma}{2})$ by assumption. Applying Theorem 6.1.11 completes the proof. $\square$

6.2 A Bayes Formula for Non-linear Filtering with Gaussian and Cox Noise

It is known [104] that if the observation noise is Brownian, the Zakai equation is a stochastic partial differential equation (SPDE) driven by a Brownian motion. Motivated by attempts to solve this equation, [56, 80] initiated the study of SPDEs driven by Brownian motion. Our purpose in this section is to show that if the noise is a Lévy process we get an SPDE driven by a Lévy process for the unconditional density. In order to include both the Gaussian and the non-Gaussian part, we shall present here the recent work from [64].

6.2.1 Introduction to the Problem

The general filtering setting can be described as follows. Assume a partially observable process $(X, Y) = (X_t, Y_t)_{0 \le t \le T} \in \mathbb{R}^2$ defined on a probability space $(\Omega, \mathcal{F}, \mathbb{P})$. The real valued process X_t denotes the unobservable component, referred to as the *signal process* or *system process*, whereas Y_t is the observable part, called the *observation process*. Thus information about X_t can only be obtained by extracting the information about X that is contained in the observation Y_t in the best possible way. In filter theory this is done by determining the conditional distribution of X_t given the information σ-field $\mathcal{F}_t^Y$ generated by $Y_s, 0 \le s \le t$. Or stated in an equivalent way, the objective is to compute the optimal filter as the conditional expectation

$$\mathbb{E}_{\mathbb{P}}[f(X_t) \mid \mathcal{F}_t^Y]$$

for a rich enough class of functions f.

In the classical non-linear filter setting, the dynamics of the observation process Y_t is supposed to follow the following Itô process

$$dY_t = h(t, X_t)dt + dW_t,$$

where W_t is a Brownian motion independent of X. Under certain conditions on the drift $h(t, X_t)$ see [51, 52], Kallianpur and Striebel derived a Bayes type formula for the conditional distribution expressed in terms of the so-called unnormalized conditional distribution. In the special case when the dynamics of the signal follows an Itô diffusion

$$dY_t = b(t, X_t)dt + \sigma(t, X_t)dB_t,$$

for a second Brownian motion B_t, Zakai [104] showed under certain conditions that the unnormalized conditional density is the solution of an associated stochastic partial differential equation, the so-called *Zakai equation*.

Here we extend the classical filter model to the following more general setting. For a general signal process X we suppose the observation model is given as

$$Y_t = \beta(t, X) + G_t + \int_0^t \int_{\mathbb{R}_0} \varsigma\, N_\lambda(dt, d\varsigma), \tag{6.2.1}$$

where

- G_t is a general Gaussian process with zero mean and continuous covariance function $R(s, t), 0 \leq s, t \leq T$, that is, independent of the signal process X.
- Let $\mathcal{F}_t^Y$ (respectively $\mathcal{F}_t^X$) denote the σ-algebra generated by $\{Y_s, 0 \leq s \leq t\}$ (respectively $\{X_s, 0 \leq s \leq t\}$) augmented by the null-sets. Define the filtration $(\mathcal{F}_t)_{0 \leq t \leq T}$ through $\mathcal{F}_t := \mathcal{F}_T^X \vee \mathcal{F}_t^Y$. Then we assume that the process

$$L_t := \int_0^t \int_{\mathbb{R}_0} \varsigma N_\lambda(dt, d\varsigma)$$

is a pure jump $\mathcal{F}_t$-semimartingale determined through the integer valued random measure N_λ that has an $\mathcal{F}_t$-predictable compensator of the form

$$\mu(dt, d\varsigma, \omega) = \lambda(t, X, \varsigma) dt \nu(d\varsigma)$$

for a Lévy measure ν and a functional $\lambda(t, X(\omega), \varsigma)$. In particular, G_t and L_t are independent.
- The function $\beta : [0, T] \times \mathbb{R}^{[0,T]} \to \mathbb{R}$ is such that $\beta(t, \cdot)$ is $\mathcal{F}_t^X$-measurable and $\beta(\cdot, X(\omega))$ is in $H(R)$ for almost all ω, where $H(R)$ denotes the Hilbert space generated by $R(s, t)$ (see Sect. 6.2.2).

The observation dynamics thus consists of an information drift of the signal disturbed by some Gaussian noise plus a pure jump part whose jump intensity depends on the signal. Note that a jump process of the form given above is also referred to as a *Cox process*.

The objective here is a first step toward extending the Kallianpur–Striebel Bayes type formula to the generalized filter setting described above. When there are no jumps present in the observation dynamics (6.2.1), the corresponding formula has been developed in [62]. We will extend their approach to the present setting including Cox noise.

In a second step we then derive a Zakai type measure-valued stochastic differential equation for the unnormalized conditional distribution of the filter. For this purpose we assume the signal process X to be a Markov process with generator $\mathcal{O}_t := \mathcal{L}_t + \mathcal{B}_t$ given as

$$\mathcal{L}_t f(x) := b(t, x)\, \partial_x f(x) + \frac{1}{2}\sigma^2(t, x)\, \partial_{xx} f(x),$$

$$\mathcal{B}_t f(x) := \int_{\mathbb{R}_0} \{f(x + \gamma(t, x)\varsigma) - f(x) - \partial_x f(x)\gamma(t, x)\varsigma\}\, \upsilon(d\varsigma)$$

where the coefficients $b(t, x)$, $\sigma(t, x)$, $\gamma(t, x)$ and $f(x)$ are in $C_0^2(\mathbb{R})$ for every t. Here, $C_0^2(\mathbb{R})$ is the space of continuous functions with compact support and bounded derivatives up to order 2. Further, we develop a Zakai type stochastic parabolic integro-partial differential equation for the unnormalized conditional density, given it exists. In the case when the dynamics of X does not contain any jumps and the Gaussian noise G_t in the observation is Brownian, the corresponding Zakai equation has also been studied in [75]. For further information on Zakai equations in a semimartingale setting we also refer to [34, 37].

The remaining part of this chapter is organized as follows. In Sect. 6.2.2 we briefly recall some theory of reproducing kernel Hilbert spaces. In Sect. 6.2.3 we obtain the Kallianpur–Striebel formula, before we discuss the Zakai type equations in Sect. 6.2.4.

6.2.2 Reproducing Kernel Hilbert Space and Stochastic Processes

A Hilbert space H consisting of real valued functions on some set $\mathbb{T}$ is said to be a *reproducing kernel Hilbert space* (RKHS) if there exists a function K on $\mathbb{T} \times \mathbb{T}$ with the following two properties: for every t in $\mathbb{T}$ and g in H,

(i) $K(\cdot, t) \in H$,
(ii) $(g(\cdot), K(\cdot, t)) = g(t)$ (the reproducing property).

K is called the *reproducing kernel* of H. The following basic properties can be found in [7].

(1) If a reproducing kernel exists, then it is unique.
(2) If K is the reproducing kernel of a Hilbert space H, then $\{K(\cdot, t), t \in \mathbb{T}\}$ spans H.
(3) If K is the reproducing kernel of a Hilbert space H, then it is nonnegative definite in the sense that for all $t_1, \ldots, t_n$ in $\mathbb{T}$ and $a_1, \ldots, a_n \in \mathbb{R}$

$$\sum_{i,j=1}^{n} K(t_i, t_j) a_i a_j \geq 0.$$

The converse of (3), stated in Theorem 6.2.1 below, is a fundamental step towards understanding the RKHS representation of Gaussian processes. A proof of the theorem can be found in [7].

Theorem 6.2.1 (E.H. Moore) *A symmetric nonnegative definite function K on $\mathbb{T} \times \mathbb{T}$ generates a unique Hilbert space, which we denote by $H(K)$ or sometimes by $H(K, \mathbb{T})$, of which K is the reproducing kernel.*

Now suppose $K(s, t)$, $s, t \in \mathbb{T}$, is a nonnegative definite function. Then, by Theorem 6.2.1, there is a RKHS, $H(K, \mathbb{T})$, with K as its reproducing kernel. If we restrict K to $\mathbb{T}' \times \mathbb{T}'$ where $\mathbb{T}' \subset \mathbb{T}$, then K is still a nonnegative definite function.

Hence K restricted to $\mathbb{T}' \times \mathbb{T}'$ will also correspond to a reproducing kernel Hilbert space $H(K, \mathbb{T}')$ of functions defined on $\mathbb{T}'$. The following result from [7; p. 351] explains the relationship between these two.

Theorem 6.2.2 *Suppose $K_{\mathbb{T}}$, defined on $\mathbb{T} \times \mathbb{T}$, is the reproducing kernel of the Hilbert space $H(K_{\mathbb{T}})$ with the norm $\| \cdot \|$. Let $\mathbb{T}' \subset \mathbb{T}$ and $K_{\mathbb{T}'}$ be the restriction of $K_{\mathbb{T}}$ on $\mathbb{T}' \times \mathbb{T}'$. Then $H(K_{\mathbb{T}'})$ consists of all f in $H(K_{\mathbb{T}})$ restricted to $\mathbb{T}'$. Further, for such a restriction $f' \in H(K_{\mathbb{T}'})$ the norm $\|f'\|_{H(K_{\mathbb{T}'})}$ is the minimum of $\|f\|_{H(K_{\mathbb{T}})}$ for all $f \in H(K_{\mathbb{T}})$ whose restriction to $\mathbb{T}'$ is f'.*

If $K(s, t)$ is the covariance function for some zero mean process $Z_t, t \in \mathbb{T}$, then, by Theorem 6.2.1, there exists a unique RKHS, $H(K, \mathbb{T})$, for which K is the reproducing kernel. It is also easy to see [e.g., see Theorem 3D, 81] that there exists a congruence (linear, one-to-one, inner product preserving map) between $H(K)$ and $\overline{sp}^{L^2}\{Z_t, t \in \mathbb{T}\}$ which takes $K(\cdot, t)$ to Z_t. Let us denote by $\langle Z, h \rangle \in \overline{sp}^{L^2}\{Z_t, t \in \mathbb{T}\}$ the image of $h \in H(K, \mathbb{T})$ under the congruence.

We conclude this section with an important special case. Suppose the stochastic process Z_t is a Gaussian process given by

$$Z_t = \int_0^t F(t, u)dW_u, \quad 0 \le t \le T$$

where $\int_0^t F^2(t, u)du < \infty$ for all $0 \le t \le T$ and W_u is Brownian motion. Then the covariance function

$$K(s, t) \equiv E(Z_s Z_t) = \int_0^{t \wedge s} F(t, u)F(s, u)du \tag{6.2.2}$$

and the corresponding RKHS is given by

$$H(K) = \left\{ g : g(t) = \int_0^t F(t, u)g^*(u)du, \quad 0 \le t \le T \right\} \tag{6.2.3}$$

for some (necessarily unique)

$$g^* \in \overline{sp}^{L^2}\{F(t, \cdot)1_{[0, t]}(\cdot), 0 \le t \le T\}$$

with the inner product

$$(g_1, g_2)_{H(K)} = \int_0^T g_1^*(u)g_2^*(u)du,$$

where

$$g_1(s) = \int_0^s F(s, u)g_1^*(u)du \quad \text{and} \quad g_2(s) = \int_0^s F(s, u)g_2^*(u)du.$$

For $0 \leq t \leq T$, by taking $K(\cdot, t)^*$ to be $F(t, \cdot)1_{[0,t]}(\cdot)$, we see, from (6.2.2) and (6.2.3), that $K(\cdot, t) \in H(K)$. To check the reproducing property suppose $h(t) = \int_0^t F(t, u)h^*(u)\, du \in H(K)$. Then

$$(h, K(\cdot, t))_{H(K)} = \int_0^T h^*(u) K(\cdot, t)^* du = \int_0^t h^*(u) F(t, u) du = h(t).$$

It is also very easy to check in this case [cf. 82, Theorem 4D] that the congruence between $H(K)$ and $\overline{sp}^{L^2}\{Z_t, t \in \mathbb{T}\}$ is given by

$$\langle Z, g \rangle = \int_0^T g^*(u) dW_u. \tag{6.2.4}$$

6.2.3 The Filter Setting and a Bayes Formula

Assume a partially observable process $(X, Y) = (X_t, Y_t)_{0 \leq t \leq T} \in \mathbb{R}^2$ defined on a probability space $(\Omega, \mathcal{F}, \mathbb{P})$. The real valued process X_t denotes the unobservable component, referred to as the *signal process*, whereas Y_t is the observable part, called the *observation process*. In particular, we assume that the dynamics of the observation process is given as follows:

$$Y_t = \beta(t, X) + G_t + \int_0^t \int_{\mathbb{R}_0} \varsigma N_\lambda(dt, d\varsigma), \tag{6.2.5}$$

where

- G_t is a Gaussian process with zero mean and continuous covariance function $R(s, t), 0 \leq s, t \leq T$, that is, independent of the signal process X.
- The function $\beta : [0, T] \times \mathbb{R}^{[0,T]} \rightarrow \mathbb{R}$ is such that $\beta(t, \cdot)$ is $\mathcal{F}_t^X$-measurable and $\beta(\cdot, X(\omega))$ is in $H(R)$ for almost all ω, where $H(R)$ denotes the Hilbert space generated by $R(s, t)$ (see Sect. 6.2.2).
- Let $\mathcal{F}_t^Y$ (respectively $\mathcal{F}_t^X$) denote the σ-algebra generated by $\{Y_s, 0 \leq s \leq t\}$ (respectively $\{X_s, 0 \leq s \leq t\}$) augmented by the null-sets. Define the filtration $(\mathcal{F}_t)_{0 \leq t \leq T}$ through $\mathcal{F}_t := \mathcal{F}_T^X \vee \mathcal{F}_t^Y$. Then we assume that the process

$$L_t := \int_0^t \int_{\mathbb{R}_0} \varsigma N_\lambda(dt, d\varsigma)$$

is a pure jump $\mathcal{F}_t$-semimartingale determined through the integer valued random measure N_λ that has an $\mathcal{F}_t$-predictable compensator of the form

$$\mu(dt, d\varsigma, \omega) = \lambda(t, X, \varsigma)dt\nu(d\varsigma)$$

for a Lévy measure ν and a functional $\lambda(t, X(\omega), \varsigma)$.

- The functional $\lambda(t, X, \varsigma)$ is assumed to be strictly positive and such that

$$\int_0^T \int_{\mathbb{R}_0} \log^2 (\lambda(s, X, \varsigma)) \, \mu(ds, d\varsigma) < \infty \quad \text{a.s.} \tag{6.2.6}$$

$$\int_0^T \int_{\mathbb{R}_0} \log^2 (\lambda(s, X, \varsigma)) \, ds \, \nu(d\varsigma) < \infty \quad \text{a.s.} \tag{6.2.7}$$

and

$$\Lambda_t := \exp \left\{ \int_0^t \int_{\mathbb{R}_0} \log \left(\frac{1}{\lambda(s, X, \varsigma)} \right) \tilde{N}_\lambda(ds, d\varsigma) \right. $$
$$\left. + \int_0^t \int_{\mathbb{R}_0} \left(\log \left(\frac{1}{\lambda(s, X, \varsigma)} \right) - \frac{1}{\lambda(s, X, \varsigma)} + 1 \right) \mu(ds, d\varsigma) \right\}$$

is a well-defined $\mathcal{F}_t$-martingale. Here $\tilde{N}_\lambda(ds, d\varsigma)$ is the compensated jump measure

$$\tilde{N}_\lambda(ds, d\varsigma) := N_\lambda(ds, d\varsigma) - \mu(dt, d\varsigma).$$

Remark 6.2.3 Note that the specific form of the predictable compensator $\mu(dt, d\varsigma, \omega)$ implies that L_t is a process with conditionally independent increments with respect to the σ-algebra $\mathcal{F}_T^X$, i.e.

$$\mathbb{E}_{\mathbb{P}}[f(L_t - L_s)1_A \mid \mathcal{F}_T^X] = \mathbb{E}_{\mathbb{P}}[f(L_t - L_s) \mid \mathcal{F}_T^X] \mathbb{E}_{\mathbb{P}}[1_A \mid \mathcal{F}_T^X],$$

for all bounded measurable functions f, $A \in \mathcal{F}_s$, and $0 \le s < t \le T$ (see, for example, Theorem 6.6 in [48]). It also follows that the process G is independent from the random measure $N_\lambda(ds, d\varsigma)$.

Given a Borel measurable function f, our non-linear filtering problem then comes down to determining the least squares estimate of $f(X_t)$, given the observations up to time t. In other words, the problem consists in evaluating the *optimal filter*

$$\mathbb{E}_{\mathbb{P}}[f(X_t) \mid \mathcal{F}_t^Y]. \tag{6.2.8}$$

In this section we want to derive a Bayes formula for the optimal filter (6.2.8) by an extension of the reference measure method presented in [62] for the purely Gaussian case. For this purpose, define for each $0 \le t \le T$ with $\beta(\cdot) = \beta(\cdot, X)$

$$\Lambda_t' := \exp \left\{ -\langle G, \beta \rangle_t - \frac{1}{2} \|\beta\|_t^2 \right\}.$$

Then the main tool is the following extension of Theorem 3.1 in [62].

Lemma 6.2.4 *Define*

$$dQ := \Lambda_t \Lambda_t' \, dP.$$

Then Q_t *is a probability measure, and under* Q_t *we have that*

$$Y_t = \widetilde{G}_t + L_t,$$

where $\widetilde{G}_s = \beta(s, X) + G_s$, $0 \leq s \leq t$, *is a Gaussian process with zero mean and covariance function* R, L_s, $0 \leq s \leq t$, *is a pure jump Lévy process with Lévy measure* ν, *and the process* X_s, $0 \leq s \leq T$ *has the same distribution as under* P. *Further, the processes* $\widetilde{G}$, L *and* X *are independent under* Q_t.

Proof Fix $0 \leq t \leq T$. First note that since $\beta(\cdot) \in H(R)$ almost surely, we have by Theorem 6.2.2 that $\beta|_{[0,t]} \in H(R; t)$ almost surely. Further, by the independence of the Gaussian process G from X and from the random measure $N_\lambda(ds, d\varsigma)$ it follows that

$$\mathbb{E}_P[\Lambda_t \Lambda_t'] = \mathbb{E}_P[\mathbb{E}_P[\Lambda_t \mid \mathcal{F}_T^X] \mathbb{E}_P[\Lambda_t' \mid \mathcal{F}_T^X]].$$

Since for $f \in H(R; t)$ the random variable $\langle G, f \rangle_t$ is Gaussian with zero mean and variance $\| f \|_t^2$, it follows again by the independence of G from X and the martingale property of Λ_t that $\mathbb{E}_P[\Lambda_t \Lambda_t'] = 1$, and Q_t is a probability measure.

Now take $0 \leq s_1, \ldots, s_m \leq t$, $0 \leq r_1, \ldots, r_p \leq t$, $0 \leq t_1, \ldots, t_n \leq T$ and real numbers $\lambda_1, \ldots, \lambda_m$, $\gamma_1, \ldots, \gamma_p$, $\alpha_1, \ldots, \alpha_n$ and consider the joint characteristic function

$$\mathbb{E}_{Q_t} \left[e^{i \sum_{j=1}^n \alpha_j X_{t_j} + i \sum_{i=1}^m \lambda_i \widetilde{G}_{s_i} + i \sum_{k=1}^p \gamma_k (L_{r_k} - L_{r_{k-1}})} \right]$$

$$= \mathbb{E}_P \left[e^{i \sum_{j=1}^n \alpha_j X_{t_j} + i \sum_{i=1}^m \lambda_i \widetilde{G}_{s_i} + i \sum_{k=1}^p \gamma_k (L_{r_k} - L_{r_{k-1}})} \Lambda_t \Lambda_t' \right]$$

$$= \mathbb{E}_P \left[e^{i \sum_{j=1}^n \alpha_j X_{t_j}} \mathbb{E}_P[e^{i \sum_{i=1}^m \lambda_i \widetilde{G}_{s_i}} \Lambda_t' \mid \mathcal{F}_T^X] \mathbb{E}_P[e^{i \sum_{k=1}^p \gamma_k (L_{r_k} - L_{r_{k-1}})} \Lambda_t \mid \mathcal{F}_T^X] \right].$$

Here, for computational convenience, the part of the characteristic function that concerns L is formulated in terms of increments of L (where we set $r_0 = 0$). Now, as in Theorem 3.1 in [62], we get by the independence of G from X that

$$\mathbb{E}_P[e^{i \sum_{i=1}^m \lambda_i \widetilde{G}_{s_i}} \Lambda_t' \mid \mathcal{F}_T^X] = e^{-\sum_{i,l=1}^m \lambda_i \lambda_l R(s_i, s_l)},$$

which is the characteristic function of a Gaussian process with mean zero and covariance function R.

Further, by the conditional independent increments of L, as in the proof of Theorem 6.6 in [48], we get that

$$\mathbb{E}_{\mathbb{P}}\left[e^{\int_r^u \int_{\mathbb{R}_0} \delta(s,X,\varsigma)\,\tilde{N}_\lambda(ds,d\varsigma)} \mid \mathcal{F}_T^X\right] = e^{\int_r^u \int_{\mathbb{R}_0} \left(e^{\delta(s,X,\varsigma)} - 1 - \delta(s,X,\varsigma)\right)\mu(dt,d\varsigma)}$$

for $0 \le r \le u \le T$. So that for one increment one obtains

$$\mathbb{E}_{\mathbb{P}}\left[e^{i\gamma(L_u - L_r)}\Lambda_t \mid \mathcal{F}_T^X\right]$$

$$= \mathbb{E}_{\mathbb{P}}\left[\exp\left\{\int_r^u \int_{\mathbb{R}_0} \left(i\gamma\varsigma + \log\left(\frac{1}{\lambda(s,X,\varsigma)}\right)\right)\tilde{N}_\lambda(ds,d\varsigma)\right.\right.$$

$$\left.\left. + \int_r^u \int_{\mathbb{R}_0}\left(i\gamma\varsigma + \log\left(\frac{1}{\lambda(s,X,\varsigma)}\right) - \frac{1}{\lambda(s,X,\varsigma)} + 1\right)\mu(dt,d\varsigma)\right\} \mid \mathcal{F}_T^X\right]$$

$$= \mathbb{E}_{\mathbb{P}}\left[\exp\left\{\int_r^u \int_{\mathbb{R}_0}\left(e^{i\gamma\varsigma + \log\left(\frac{1}{\lambda(s,X,\varsigma)}\right)} - 1 - i\gamma\varsigma - \log\left(\frac{1}{\lambda(s,X,\varsigma)}\right)\right)\mu(dt,d\varsigma)\right.\right.$$

$$\left.\left. + \int_r^u \int_{\mathbb{R}_0}\left(i\gamma\varsigma + \log\left(\frac{1}{\lambda(s,X,\varsigma)}\right) - \frac{1}{\lambda(s,X,\varsigma)} + 1\right)\mu(dt,d\varsigma)\right\} \mid \mathcal{F}_T^X\right]$$

$$= \mathbb{E}_{\mathbb{P}}\left[\exp\left\{\int_r^u \int_{\mathbb{R}_0}\left(e^{i\gamma\varsigma + \log\left(\frac{1}{\lambda(s,X,\varsigma)}\right)} - \frac{1}{\lambda(s,X,\varsigma)}\right)\lambda(t,X,\varsigma)dt\nu(d\varsigma)\right\} \mid \mathcal{F}_T^X\right]$$

$$= \exp\left\{(u-r)\int_{\mathbb{R}_0}\left(e^{i\gamma\varsigma} - 1\right)\nu(d\varsigma)\right\}.$$

The generalization to the sum of increments is straightforward and one obtains the characteristic function of the finite dimensional distribution of a Lévy process (of finite variation):

$$\mathbb{E}_{\mathbb{P}}[e^{i\sum_{k=1}^p \gamma_k(L_{r_k} - L_{r_{k-1}})}\Lambda_t \mid \mathcal{F}_T^X] = \exp\left\{\sum_{k=1}^p (r_k - r_{k-1})\int_{\mathbb{R}_0}\left(e^{i\gamma_k\varsigma} - 1\right)\nu(d\varsigma)\right\}.$$

All together we end up with

$$\mathbb{E}_{\mathbb{Q}_t}\left[e^{i\sum_{j=1}^n \alpha_j X_{t_j} + i\sum_{i=1}^m \lambda_i \tilde{G}_{s_i} + i\sum_{k=1}^p \gamma_k(L_{r_k} - L_{r_{k-1}})}\right]$$

$$= \mathbb{E}_{\mathbb{P}}\left[e^{i\sum_{j=1}^n \alpha_j X_{t_j}}\right] \cdot e^{-\sum_{i,l=1}^m \lambda_i\lambda_l R(s_i,s_l)} \cdot e^{\sum_{k=1}^p (r_k - r_{k-1})\int_{\mathbb{R}_0}(e^{i\gamma_k\varsigma} - 1)\nu(d\varsigma)},$$

which completes the proof.   □

Remark 6.2.5 Note that in the case where G is Brownian motion Lemma 6.2.4 is just the usual Girsanov theorem for Brownian motion and random measures. In this case, it follows from the Cameron–Martin theorem and the fact that X is independent of G, that $\Lambda_t \Lambda_t'$ is a martingale, and $d\mathbb{Q}$ is a probability measure.

Now, the inverse Radon–Nikodym derivative

$$\frac{d\mathbb{P}}{d\mathbb{Q}_t} = (\Lambda_t)^{-1}(\Lambda_t')^{-1}$$

is $\mathbb{Q}_t$–a.s. by condition (6.2.6) and an argument as in [62, p. 857] via

$$(\Lambda_t)^{-1} = \exp\left\{ \int_0^t \int_{\mathbb{R}_0} \log\left(\lambda(s, \dot{X}, \varsigma)\right) \widetilde{N}(ds, d\varsigma) \right.$$
$$\left. + \int_0^t \int_{\mathbb{R}_0} \left(\log\left(\lambda(s, X, \varsigma)\right) - \lambda(s, X, \varsigma) + 1\right) ds\, \nu(d\varsigma) \right\},$$
$$(\Lambda_t')^{-1} = \exp\left\{ \langle \widetilde{G}, \beta \rangle_t - \frac{1}{2}\|\beta\|_t^2 \right\}.$$

Here

$$\widetilde{N}(ds, d\varsigma) := N_\lambda(ds, d\varsigma) - dt\nu(d\varsigma)$$

is now a compensated Poisson random measure under $\mathbb{Q}_t$. Then we have by the Bayes formula for conditional expectation for any $\mathcal{F}_T^X$-measurable integrable function $g(T, X)$

$$\mathbb{E}_\mathbb{P}\left[g(T, X) \,|\, \mathcal{F}_t^Y \right] = \frac{\mathbb{E}_{\mathbb{Q}_t}\left[g(T, X)(\Lambda_t)^{-1}(\Lambda_t')^{-1} \,|\, \mathcal{F}_t^Y \right]}{\mathbb{E}_{\mathbb{Q}_t}\left[(\Lambda_t)^{-1}(\Lambda_t')^{-1} \,|\, \mathcal{F}_t^Y \right]}.$$

From Lemma 6.2.4 we know that the processes $(\widetilde{G}_s)_{0 \le s \le t}$, $(L_s)_{0 \le s \le t}$, and $(X_s)_{0 \le s \le T}$ are independent under $\mathbb{Q}_t$ and that the distribution of X is the same under $\mathbb{Q}_t$ as under $\mathbb{P}$. Hence conditional expectations of the form $E_{\mathbb{Q}_t}[\phi(X, \widetilde{G}, L) \,|\, \mathcal{F}_t^Y]$ can be computed as

$$\mathbb{E}_{\mathbb{Q}_t}[\phi(X, \widetilde{G}, L) \,|\, \mathcal{F}_t^Y](\omega) = \int_\Omega \phi(X(\hat{\omega}), \widetilde{G}(\omega), L(\omega))\mathbb{Q}_t(d\hat{\omega})$$
$$= \int_\Omega \phi(X(\hat{\omega}), \widetilde{G}(\omega), L(\omega))\mathbb{P}(d\hat{\omega})$$
$$= \mathbb{E}_{\hat{\mathbb{P}}}[\phi(X(\hat{\omega}), \widetilde{G}(\omega), L(\omega))],$$

where $(\omega, \hat{\omega}) \in \Omega \times \Omega$ and the index $\hat{\mathbb{P}}$ denotes integration with respect to $\hat{\omega}$. Consequently, we get the following Bayes formula for the optimal filter.

Theorem 6.2.6 *Under the above specified conditions, for any $\mathcal{F}_T^X$-measurable integrable function $g(T, X)$*

$$\mathbb{E}_\mathbb{P}\left[g(T, X) \,|\, \mathcal{F}_t^Y \right] = \frac{\int_\Omega g(T, X(\hat{\omega}))\alpha_t(\omega, \hat{\omega})\alpha_t'(\omega, \hat{\omega})\mathbb{P}(d\hat{\omega})}{\int_\Omega \alpha_t(\omega, \hat{\omega})\alpha_t'(\omega, \hat{\omega})\mathbb{P}(d\hat{\omega})}$$
$$= \frac{\mathbb{E}_{\hat{\mathbb{P}}}\left[g(T, X(\hat{\omega}))\alpha_t(\omega, \hat{\omega})\alpha_t'(\omega, \hat{\omega}) \right]}{\mathbb{E}_{\hat{\mathbb{P}}}\left[\alpha_t(\omega, \hat{\omega})\alpha_t'(\omega, \hat{\omega}) \right]},$$

where

$$\alpha_t(\omega, \hat{\omega}) = \exp \left\{ \int_0^t \int_{\mathbb{R}_0} \log \left(\lambda(s, X(\hat{\omega}), \varsigma) \right) \tilde{N}(\omega, ds, d\varsigma) \right.$$

$$\left. + \int_0^t \int_{\mathbb{R}_0} \left(\log \left(\lambda(s, X(\hat{\omega}), \varsigma) \right) - \lambda(s, X(\hat{\omega}), \varsigma) + 1 \right) ds \, \nu(d\varsigma) \right\},$$

$$\alpha_t'(\omega, \hat{\omega}) = \exp \left\{ \langle \tilde{G}(\omega), \beta(\cdot, \hat{\omega}) \rangle_t - \frac{1}{2} \| \beta(\cdot, \hat{\omega}) \|_t^2 \right\}.$$

6.2.4 Zakai Type Equations

Using the Bayes formula from above we now want to proceed further in deriving a Zakai type equation for the unnormalized filter. This equation is basic in order to obtain the filter recursively. To this end we have to impose certain restrictions on both the signal process and the Gaussian part of the observation process.

Regarding the signal process X, we assume its dynamics to be Markov. More precisely, we consider the parabolic integro-differential operator $\mathcal{O}_t := \mathcal{L}_t + \mathcal{B}_t$, where

$$\mathcal{L}_t f(x) := b(t, x) \, \partial_x f(x) + \frac{1}{2} \sigma^2(t, x) \, \partial_{xx} f(x),$$

$$\mathcal{B}_t f(x) := \int_{\mathbb{R}_0} \left\{ f(x + \gamma(t, x)\varsigma) - f(x) - \partial_x f(x) \gamma(t, x)\varsigma \right\} \upsilon(d\varsigma)$$

for $f \in C_0^2(\mathbb{R})$. Here, $C_0^2(\mathbb{R})$ is the space of continuous functions with compact support and bounded derivatives up to order 2. Further, we suppose that $b(t, \cdot)$, $\sigma(t, \cdot)$, and $\gamma(t, \cdot)$ are in $C_0^2(\mathbb{R})$ for every t and that $\upsilon(d\varsigma)$ is a Lévy measure with second moment. The signal process X_t, $0 \le t \le T$, is then assumed to be a solution of the martingale problem corresponding to $\mathcal{O}_t$, i.e.

$$f(X_t) - \int_0^t (\mathcal{O}_u f)(X_u) du$$

is an $\mathcal{F}_t^X$-martingale with respect to $\mathbb{P}$ for every $f \in C_0^2(\mathbb{R})$.

Further, we restrict the Gaussian process G of the observation process in (6.2.5) to belong to the special case presented in Sect. 2.1, i.e.

$$G_t = \int_0^t F(t, s) dW_s,$$

where W_t is Brownian motion and $F(t, s)$ is a deterministic function such that $\int_0^t F^2(t, s) \, ds$, $0 \le t \le T$. Note that this type of process includes both Ornstein–

Uhlenbeck processes as well as fractional Brownian motion. Then $\beta(t, X)$ will be of the form

$$\beta(t, X) = \int_0^t F(t, s)h(s, X_s)ds.$$

Further, with

$$\widetilde{W}_t := \int_0^t h(s, X_s)ds + W_t$$

we get $\langle \widetilde{G}, \beta \rangle_t = \int_0^t h(s, X_s)\, d\widetilde{W}_s$ and $\|\beta\|_t^2 = \int_0^t h^2(s, X_s)\, ds$, and $\alpha_t'(\omega, \hat{\omega})$ in Theorem 6.2.6 becomes

$$\alpha_t'(\omega, \hat{\omega}) = \exp\left\{ \int_0^t h(s, X_s(\hat{\omega}))d\widetilde{W}_s(\omega) - \frac{1}{2}\int_0^t h^2(s, X_s(\hat{\omega}))ds \right\}.$$

Note that in this case $\widetilde{W}_s$, $0 \le s \le t$, is a Brownian motion under $\mathbb{Q}_t$.
   For $f \in C_0^2(\mathbb{R})$ we now define the unnormalized filter $V_t(f) = V_t(f)(\omega)$ by

$$V_t(f)(\omega) := \int_\Omega f(X_t(\hat{\omega}))\alpha_t(\omega, \hat{\omega})\alpha_t'(\omega, \hat{\omega})\mathbb{P}(d\hat{\omega}) = \mathbb{E}_{\hat{\mathbb{P}}}\left[f(X_t(\hat{\omega}))\alpha_t(\omega, \hat{\omega})\alpha_t'(\omega, \hat{\omega})\right].$$

Then this unnormalized filter obeys the following dynamics.

Theorem 6.2.7 (*Zakai equation I*) *Under the above specified assumptions, the unnormalized filter $V_t(f)$ satisfies the equation*

$$dV_t\Big(f(\cdot)\Big)(\omega) = V_t\Big(\mathcal{O}_t f(\cdot)\Big)(\omega)dt + V_t\Big(h(t, \cdot)f(\cdot)\Big)(\omega)d\widetilde{W}_t(\omega)$$
$$+ \int_{\mathbb{R}_0} V_t\Big((\lambda(t, \cdot, \varsigma) - 1)\, f(\cdot)\Big)(\omega)\widetilde{N}(\omega, dt, d\varsigma). \qquad (6.2.9)$$

Proof Set

$$g_t(\hat{\omega}) := f(X_T(\hat{\omega})) - \int_t^T (\mathcal{O}_s f)(X_s(\hat{\omega}))ds.$$

Then, by our assumptions on the coefficients b, σ, γ and on the Lévy measure $\upsilon(d\varsigma)$, we have $|g_t| < C$ for some constant C. Since $f(X_t) - \int_0^t \mathcal{O}_t f(X_s)\, ds$ is a martingale we obtain

$$\mathbb{E}_{\hat{\mathbb{P}}}[g_t \mid \mathcal{F}_t^{X(\hat{\omega})}] = f(X_t), \quad 0 \le t \le T. \qquad (6.2.10)$$

If we define

$$\Gamma_t(\omega, \hat{\omega}) := \alpha_t(\omega, \hat{\omega})\alpha'_t(\omega, \hat{\omega}),$$

then, because $\Gamma_t(\omega, \hat{\omega})$ is $\mathcal{F}_t^{X(\hat{\omega})}$-measurable for each ω, Eq. (6.2.10) implies that

$$V_t(f) = \mathbb{E}_{\hat{\mathbb{P}}}[f(X_t(\hat{\omega}))\Gamma_t(\omega, \hat{\omega})]$$
$$= \mathbb{E}_{\hat{\mathbb{P}}}[\mathbb{E}_{\hat{\mathbb{P}}}[g_t(\hat{\omega})\Gamma_t(\omega, \hat{\omega}) \,|\, \mathcal{F}_t^{X(\hat{\omega})}]] = \mathbb{E}_{\hat{\mathbb{P}}}[g_t(\hat{\omega})\Gamma_t(\omega, \hat{\omega})].$$

By definition of g_t,

$$dg_t(\hat{\omega}) = (\mathcal{O}_t f)(X_t(\hat{\omega}))dt.$$

Furthermore, $\Gamma_t = \Gamma_t(\omega, \hat{\omega})$ is the Doléans–Dade solution of the following linear SDE

$$d\Gamma_t = h(t, X_t(\hat{\omega})\Gamma_t d\widetilde{W}_t(\omega) + \int_{\mathbb{R}_0} \left(\lambda(t, X_t(\hat{\omega}), \varsigma) - 1\right) \Gamma_t \widetilde{N}(\omega, dt, d\varsigma).$$

So we get

$$\mathbb{E}_{\hat{\mathbb{P}}}\left[g_t(\hat{\omega})\Gamma_t\right] = \mathbb{E}_{\hat{\mathbb{P}}}\left[g_0(\hat{\omega})\Gamma_0\right] + \mathbb{E}_{\hat{\mathbb{P}}}\left[\int_0^t (\mathcal{O}_s f)(X_s(\hat{\omega}))\Gamma_s ds\right]$$
$$+ \mathbb{E}_{\hat{\mathbb{P}}}\left[\int_0^t h(s, X_s(\hat{\omega})g_s(\hat{\omega})\Gamma_s \, d\widetilde{W}_s(\omega)\right]$$
$$+ \mathbb{E}_{\hat{\mathbb{P}}}\left[\int_0^t \int_{\mathbb{R}_0} \left(\lambda(s, X_s(\hat{\omega}), \varsigma) - 1\right) g_s(\hat{\omega})\Gamma_s \widetilde{N}(\omega, ds, d\varsigma)\right].$$

The first term on the right-hand side equals $f(X_0)$, and for the second one we can invoke Fubini's theorem to get

$$\mathbb{E}_{\hat{\mathbb{P}}}\left[\int_0^t (\mathcal{O}_s f)(X_s(\hat{\omega}))\Gamma_s ds\right] = \int_0^t \mathbb{E}_{\hat{\mathbb{P}}}[(\mathcal{O}_s f)(X_s(\hat{\omega}))\Gamma_s]ds$$
$$= \int_0^t V_s\left(\mathcal{O}_s f(\cdot)\right)(\omega)ds.$$

For the third term we employ the stochastic Fubini theorem for Brownian motion (see for Example 5.14 in [58]) in order to get

$$\mathbb{E}_{\hat{\mathbb{P}}}\left[\int_0^t h(s, X_s(\hat{\omega})) g_s(\hat{\omega}) \Gamma_s d\widetilde{W}_s(\omega)\right]$$

$$= \int_0^t \mathbb{E}_{\hat{\mathbb{P}}}\left[h(s, X_s(\hat{\omega})) g_s(\hat{\omega}) \Gamma_s\right] d\widetilde{W}_s(\omega)$$

$$= \int_0^t \mathbb{E}_{\hat{\mathbb{P}}}\left[h(s, X_s(\hat{\omega})) \Gamma_s \mathbb{E}_{\hat{\mathbb{P}}}\left[g_s(\hat{\omega}) \mid \mathcal{F}_s^{X(\hat{\omega})}\right]\right] d\widetilde{W}_s(\omega)$$

$$= \int_0^t \mathbb{E}_{\hat{\mathbb{P}}}\left[h(s, X_s(\hat{\omega})) \Gamma_s f(X_s(\hat{\omega}))\right] d\widetilde{W}_s(\omega)$$

$$= \int_0^t V_s\Big(h(s, \cdot) f(\cdot)\Big)(\omega) d\widetilde{W}_s(\omega).$$

Further, one easily sees that the analogue stochastic Fubini theorem for compensated Poisson random measures holds, and we get analogously for the last term

$$\mathbb{E}_{\hat{\mathbb{P}}}\left[\int_0^t \int_{\mathbb{R}_0} (\lambda(s, X_s(\hat{\omega}), \varsigma) - 1) g_s(\hat{\omega}) \Gamma_s \widetilde{N}(\omega, ds, d\varsigma)\right]$$

$$= \int_0^t \int_{\mathbb{R}_0} V_s\Big((\lambda(s, \cdot, \varsigma) - 1) f(\cdot)\Big)(\omega) \widetilde{N}(\omega, ds, d\varsigma),$$

which completes the proof.   □

If one further assumes that the filter has a so-called unnormalized conditional density $u(t, x)$ then we can derive a stochastic integro-PDE determining $u(t, x)$ which for the Brownian motion case was first established in [104] and is usually referred to as the Zakai equation.

Definition 6.2.8 We say that a process $u(t, x) = u(\omega, t, x)$ is the unnormalized conditional density of the filter if

$$V_t(f)(\omega) = \int_{\mathbb{R}} f(x) u(\omega, t, x) dx \qquad (6.2.11)$$

for all bounded continuous functions $f : \mathbb{R} \to \mathbb{R}$.

From now on we restrict the integro part $\mathcal{B}_t$ of the operator $\mathcal{O}_t$ to be the one of a pure jump Lévy process, i.e. $\gamma = 1$, and we assume the initial value $X_0(\omega)$ of the signal process to possess a density denoted by $\xi(x)$. Then the following holds:

Theorem 6.2.9 (*Zakai equation II*) *Suppose the unnormalized conditional density* $u(t, x)$ *of our filter exists. Then, provided a solution exists,* $u(t, x)$ *solves the following stochastic integro-PDE*

$$\begin{cases} du(t, x) = \mathcal{O}_t^* u(t, x) dt + h(t, x) u(t, x) d\widetilde{W}_t(\omega) \\ \qquad\qquad + \int_{\mathbb{R}_0} (\lambda(t, x, \varsigma) - 1) u(t, x) \widetilde{N}(\omega, dt, d\varsigma) \\ u(0, x) = \xi(x). \end{cases} \qquad (6.2.12)$$

Here $\mathcal{O}_t^* := \mathcal{L}_t^* + \mathcal{B}_t^*$ is the adjoint operator of $\mathcal{O}_t$ given by

$$\mathcal{L}_t^* f(x) := -\partial_x \left(b(t,x) f(x) \right) + \frac{1}{2}\partial_{xx}\left(\sigma^2(t,x)f(x)\right),$$

$$\mathcal{B}_t^* f(x) := \int_{\mathbb{R}_0} \{f(x-\varsigma) - f(x) + \partial_x f(x)\varsigma\}\, v(d\varsigma)$$

for $f \in C_0^2(\mathbb{R})$.

For sufficient conditions on the coefficients under which there exists a classical solution of (6.2.12) see for example [75]; in [74] the existence of solutions in a generalized sense of stochastic distributions is treated.

Proof By (6.2.9) and (6.2.11) we have for all $f \in C_0^\infty(\mathbb{R})$

$$\int_{\mathbb{R}} f(x)u(t,x)dx = \int_{\mathbb{R}} f(x)\xi(x)dx + \int_0^t \int_{\mathbb{R}} u(s,x)\mathcal{O}_s^* f(x)dxds$$

$$+ \int_0^t \int_{\mathbb{R}} u(s,x)h(s,x)f(x)dxd\widetilde{W}_s(\omega)$$

$$+ \int_0^t \int_{\mathbb{R}_0} \int_{\mathbb{R}} u(s,x)\left(\lambda(s,x,\varsigma) - 1\right)f(x)dx\widetilde{N}(\omega, ds, d\varsigma).$$

Now, using integration by parts, we get

$$\int_{\mathbb{R}} u(s,x)\mathcal{L}_s^* f(x)dx = \int_{\mathbb{R}} f(x)\mathcal{O}_s^* u(s,x)dx. \qquad (6.2.13)$$

Further, using integration by parts again and by substitution, we have

$$\int_{\mathbb{R}} u(s,x)\mathcal{B}_s^* f(x)dx = \int_{\mathbb{R}} f(x)\mathcal{B}_s^* u(s,x)dx. \qquad (6.2.14)$$

Fubini together with (6.2.13) and (6.2.14) then yields

$$\int_{\mathbb{R}} f(x)u(t,x)dx = \int_{\mathbb{R}} f(x)\xi(x)dx + \int_{\mathbb{R}} f(x)\left(\int_0^t \mathcal{O}_s^* u(s,x)ds\right)dx$$

$$+ \int_{\mathbb{R}} f(x)\left(\int_0^t u(s,x)h(s,x)d\widetilde{W}_s(\omega)\right)dx$$

$$+ \int_{\mathbb{R}} f(x)\left(\int_0^t \int_{\mathbb{R}_0} u(s,x)\left(\lambda(s,x,\varsigma) - 1\right)\widetilde{N}(\omega, ds, d\varsigma)\right)dx.$$

Since this holds for all $f \in C_0^\infty(\mathbb{R})$ we get (6.2.12). $\square$

6.3 Remarks and Related Literature

Kallianpur and Striebel [52] established a Bayes formula for the filter. This can be used in order to obtain the computation of the filter iteratively using SPDEs driven by a Brownian motion (Zakai equation). This motivated the study of the pioneering work of [56, 80] on SPDEs with respect to Brownian motion.

It was shown in [62] that the Brownian motion can be replaced by any Gaussian process in establishing a Bayes formula. For a subclass of Gaussian processes (including fractional Brownian motion) one can use it to obtain an analogue of the Zakai equation.

For Lévy processes such an equation was first established in [75]. The work presented here is from [64], giving a simultaneous generalization of the work in [62, 75]. The properties of reproducing kernel Hilbert spaces are taken from the basic work of [7].

For sufficient conditions under which a classical solution to (6.2.12) exists is given in [75]. The existence of a generalized solution is proven in [74].

The material in Sect. 6.1 is taken directly from the pioneering work of [32] which constructs the SPDE as an equation in an appropriate Hilbert space, motivated by [13, 31].

Chapter 7
Stability Theory for Stochastic Semilinear Equations

In this chapter we study stability of time-homogeneous stochastic partial differential equations

$$\begin{cases} dZ_t = (AZ_t + a(Z_t))dt + \int_H f(x, Z_t)q(dt, dx) \\ Z_0 = x. \end{cases} \tag{7.0.1}$$

Throughout this chapter, $(\Omega, \mathcal{F}, (\mathcal{F}_t)_{t \geq 0}, \mathbb{P})$ denotes a filtered probability space satisfying the usual conditions. Let H be a separable Hilbert space. In (7.0.1), q denotes the compensated Poisson random measure on $\mathbb{R}_+ \times E$ associated to a Poisson random measure N with compensator $dt \otimes \beta(dx)$.

7.1 Exponential Stability in the Mean Square Sense

Let $a : H \to H$ and $f : H\backslash\{0\} \times H \to H$ be functions.

Assumption 7.1.1 There exists a constant $L > 0$ such that

$$\|a(z_1) - a(z_2)\| \leq L\|z_1 - z_2\|, \tag{7.1.1}$$

$$\int_E \|f(z_1, u) - f(z_2, u)\|^2 \beta(du) \leq L\|z_1 - z_2\|^2 \tag{7.1.2}$$

for all $z_1, z_2 \in H$, and a constant $C > 0$, such that

$$\int_E \|f(z, u)\|^2 \beta(du) \leq C(1 + \|z\|^2) \quad \text{for all } z \in H. \tag{7.1.3}$$

© Springer International Publishing Switzerland 2015
V. Mandrekar and B. Rüdiger, *Stochastic Integration in Banach Spaces*,
Probability Theory and Stochastic Modelling 73, DOI 10.1007/978-3-319-12853-5_7

Remark 7.1.2 Note that

$$\|a(z)\| \le K(\|z\| + 1), \quad z \in H, \tag{7.1.4}$$

for some constant $K > 0$.

Throughout this chapter, we assume that Assumption 7.1.1 is fulfilled.

As proved in Chap. 5, under Assumption 7.1.1, for each $x \in H$ the stochastic partial differential equation (7.0.1) has a unique mild solution $(Z_t^x)_{t \ge 0}$, and it is a homogeneous Markov process satisfying $\int_0^T \mathbb{E}[\|Z_s^x\|^2] ds < \infty$ for all $T \in \mathbb{R}_+$.

Definition 7.1.3 We say that the solution of SPDE (7.0.1) is exponentially stable in the mean square sense if there exist positive real constants c, θ such that

$$\mathbb{E}[\|Z_t^x\|^2] \le c e^{-\theta t} \|x\|^2 \quad \text{for all } x \in H \text{ and } t \ge 0. \tag{7.1.5}$$

Our object in this chapter is to derive conditions for exponential stability in the mean square sense.

By Theorem 5.4.2, for each $x \in H$ the solution Z^x for (7.0.1) is a time-homogeneous Markov process.

Let D denote the Fréchét derivative.

Definition 7.1.4 The *infinitesimal generator* $\mathcal{L}$ of the solution of SPDE (7.0.1) is defined as

$$\mathcal{L}\psi(y) = D\psi(y)(Ay + a(y))$$
$$+ \int_{H \setminus \{0\}} \Big(\psi(y + f(x, y)) - \psi(y) - D\psi(y)f(x, y) \Big) \beta(dx), \quad y \in \mathcal{D}(A)$$

for any $\psi \in C_b^{2,\text{loc}}(H; \mathbb{R})$.

Lemma 7.1.5 *Let* $\phi \in C_b^{1,2,\text{loc}}(\mathbb{R}_+ \times H; H)$ *and* $x \in \mathcal{D}(A)$ *be arbitrary, and let* Z *be a strong solution for* (7.0.1) *with* $Z_0 = x$. *Then, we have*

$$\phi(t, Z_t) - \phi(0, x) = \int_0^t (\partial_s \phi(s, Z_s) + \mathcal{L}\phi(s, Z_s)) ds$$
$$+ \int_0^t \int_E (\phi(s, Z_s + f(x, Z_s)) - \phi(s, Z_s)) q(ds, dx), \quad t \ge 0.$$

Proof The statement follows from Itô's formula (Theorem 3.7.2). □

Definition 7.1.6 A function $\psi \in C_b^{2,\text{loc}}(H; \mathbb{R})$ is called a Lyapunov function for (7.0.1) if there are constants $c_i > 0$, $i = 1, 2, 3$ such that

$$c_1 \|x\|^2 \le \psi(x) \le c_3 \|x\|^2, \quad x \in H \tag{7.1.6}$$
$$\mathcal{L}\psi(x) \le -c_2 \psi(x), \quad x \in \mathcal{D}(A). \tag{7.1.7}$$

For $n \in \mathbb{N}$ with $n > \alpha$, where the constant $\alpha \in \mathbb{R}$ stems from (5.1.2), we consider the approximative system (A.3.2), and denote by $\mathcal{L}_n$ its infinitesimal generator. According to Theorem A.3.2, for each $x \in \mathcal{D}(A)$ there exists a unique strong solution $Z^{n,x}$ for (A.3.2) with $Z_0^n = x$.

Lemma 7.1.7 *Let* $\psi \in C_b^{2,\mathrm{loc}}(H; \mathbb{R})$ *be a function such that for some constants* $c_3, k_3 > 0$ *we have*

$$|\psi(x)| \leq c_3(\|x\|^2 + 1), \quad x \in H. \qquad (7.1.8)$$

Then, for all $x \in \mathcal{D}(A)$*, all* $c_2 \geq 0$ *and all* $n \in \mathbb{N}$ *such that*

$$\mathbb{E}\left[\int_0^t |e^{c_2 s}(c_2 + \mathcal{L}_n)\psi(Z_s^{n,x})|ds\right] < \infty, \quad t \geq 0 \qquad (7.1.9)$$

we have the identity

$$e^{c_2 t}\mathbb{E}[\psi(Z_t^{n,x})] - \psi(x) = \mathbb{E}\left[\int_0^t e^{c_2 s}(c_2 + \mathcal{L}_n)\psi(Z_s^{n,x})ds\right], \quad t \geq 0. \qquad (7.1.10)$$

Proof Note that the function $\phi : \mathbb{R}_+ \times H \to H$, $\phi(t, x) = e^{c_2 t}\psi(x)$ belongs to $C_b^{1,2,\mathrm{loc}}(\mathbb{R}_+ \times H; H)$. Using Itô's formula (Lemma 7.1.5) we obtain

$$e^{c_2 t}\psi(Z_t^{n,x}) - \psi(x) = \int_0^t e^{c_2 s}(c_2 + \mathcal{L}_n)\psi(Z_t^{n,x})ds + M_t, \quad t \geq 0$$

where M denotes the local martingale

$$M_t = \int_0^t \int_E e^{cs}(\psi(Z_s^{n,x} + f(x, Z_s^{n,x})) - \psi(Z_s^{n,x}))q(ds, dx), \quad t \geq 0.$$

There exists a non-decreasing sequence $(\tau_m)_{m \in \mathbb{N}}$ of stopping times such that $\tau_m \to \infty$ almost surely and M^{τ_m} is a martingale with $M_0^{\tau_m} = 0$ for each $m \in \mathbb{N}$. Therefore, we obtain

$$e^{ct}\mathbb{E}[\psi(Z_{t \wedge \tau_m}^{n,x})] - \psi(x) = \mathbb{E}\left[\int_0^{t \wedge \tau_m} e^{cs}(c + \mathcal{L}_n)\psi(Z_s^{n,x})ds\right], \quad t \geq 0.$$

Let $t \geq 0$ be arbitrary. By (7.1.8), we have $\mathbb{P}$-almost surely

$$\psi(Z_{t \wedge \tau_m}^{n,x}) \leq c_3(\|Z_{t \wedge \tau_m}^{n,x}\|^2 + 1) \leq c_3\left(\sup_{s \in [0,t]} \|Z_s^{n,x}\|^2 + 1\right), \quad m \in \mathbb{N}.$$

Since $(Z_s)_{s \in [0,t]} \in \mathcal{S}_t^2(H)$ and (7.1.9) is valid, Lebesgue's dominated convergence theorem applies and yields (7.1.10). $\qquad \square$

At this point, we prepare a generalized version of Lebesgue's dominated convergence theorem, which we will require in the sequel.

Lemma 7.1.8 *Let $(X, \mathcal{X}, \mu)$ be a measure space. Let $f_n : X \to \mathbb{R}$, $n \in \mathbb{N}$ and $f : X \to \mathbb{R}$ be measurable such that $f_n \to f$ almost surely. Furthermore, let $g_n \in L^1(X)$, $n \in \mathbb{N}$ and $g \in L^1(X)$ be such that $g = \liminf_{n\to\infty} g_m$ almost surely, $|f_n| \leq g_n$ for all $n \in \mathbb{N}$ and $\int_X g_n d\mu \to \int_X g d\mu$. Then we have $f \in L^1(X)$ and*

$$\int_X f_n d\mu \to \int_X f d\mu.$$

Proof The hypothesis $|f_n| \leq g_n$ implies that the functions $g_n + f_n$ and $g_n - f_n$ are nonnegative and measurable for all $n \in \mathbb{N}$. Fatou's Lemma yields

$$\int_X g d\mu + \int_X f d\mu = \int_X \liminf_{n\to\infty}(g_n + f_n)d\mu \leq \liminf_{n\to\infty}\int_X (g_n + f_n)d\mu$$

$$= \int_X g d\mu + \liminf_{n\to\infty}\int_X f_n d\mu$$

as well as

$$\int_X g d\mu - \int_X f d\mu = \int_X \liminf_{n\to\infty}(g_n - f_n)d\mu \leq \liminf_{n\to\infty}\int_X (g_n - f_n)d\mu$$

$$= \int_X g d\mu - \limsup_{n\to\infty}\int_X f_n d\mu.$$

Since $g \in L^1(X)$, we deduce that

$$\int_X f d\mu \leq \liminf_{n\to\infty}\int_X f_n d\mu \leq \limsup_{n\to\infty}\int_X f_n d\mu \leq \int_X f d\mu,$$

and hence the desired conclusion follows.   □

Lemma 7.1.9 *Let $\psi \in C_b^{2,\mathrm{loc}}(H; \mathbb{R})$ be arbitrary. Suppose there exist constants $c_3, c_4, c_5 > 0$ such that we have (7.1.8) and*

$$\|D\psi(x)\| \leq c_4(1 + \|x\|), \quad x \in H \tag{7.1.11}$$

$$\|D\psi^2(x)\| \leq c_5, \quad x \in H. \tag{7.1.12}$$

Let $x \in H$, $t \geq 0$ and $c_2 > 0$ be arbitrary. Then there exists a subsequence $(n_k)_{k\in\mathbb{N}}$ such that

$$\mathbb{E}\left[\int_0^t |e^{c_2 s}(\mathcal{L}_{n_k} - \mathcal{L})\psi(Z_s^{n_k,x})|ds\right] < \infty, \quad k \in \mathbb{N} \tag{7.1.13}$$

and we have the convergences

$$\mathbb{E}[\psi(Z_t^{n_k})] \to \mathbb{E}[\psi(Z_t)], \tag{7.1.14}$$

$$\mathbb{E}\left[\int_0^t e^{c_2 s}(\mathcal{L}_{n_k} - \mathcal{L})\psi(Z_s^{n_k,x})ds\right] \to 0. \tag{7.1.15}$$

Proof Let $x \in H$, $t \geq 0$ and $c_2 > 0$ be arbitrary. By Theorem A.3.2 we have

$$\mathbb{E}\left[\sup_{s \in [0,t]} \|Z_s^{n,x} - Z_s^x\|^2\right] \to 0. \tag{7.1.16}$$

Hence, there exists a subsequence $(n_k)_{k \in \mathbb{N}}$ such that almost surely

$$\sup_{s \in [0,t]} \|Z_s^{n_k,x} - Z_s^x\|^2 \to 0. \tag{7.1.17}$$

By (7.1.8) and the generalized Lebesgue dominated convergence theorem (Lemma 7.1.8) we deduce (7.1.14). Note that for each $k \in \mathbb{N}$ we have

$$e^{c_2 s}(\mathcal{L}_{n_k} - \mathcal{L})\psi(Z_s^{n_k,x}) = X_s^k + \int_E F^k(s,u)\beta(du), \quad s \in [0,t]$$

where we have defined $X^k : \Omega \times [0,t] \to \mathbb{R}$ and $F^k : \Omega \times [0,t] \times E \to \mathbb{R}$ as

$$X_s^k = e^{c_2 s} D\psi(Z_s^{n_k,x})(n_k R_{n_k}(A) - \mathrm{Id})a(Z_s^{n_k,x}),$$

$$F^k(s,u) = e^{c_2 s}\left(\psi(Z_s^{n_k,x} + n_k R_{n_k}(A)f(u, Z_s^{n_k,x})) - \psi(Z_s^{n_k,x})\right.$$
$$\left. - D\psi(Z_s^{n_k,x})n_k R_{n_k}(A)f(u, Z_s^{n_k,x})\right)$$
$$- e^{c_2 s}\left(\psi(Z_s^{n_k,x} + f(u, Z_s^{n_k,x}))\right.$$
$$\left. - \psi(Z_s^{n_k,x}) - D\psi(Z_s^{n_k,x})f(u, Z_s^{n_k,x})\right).$$

By (7.1.17) we get

$$X_s^k \to 0, \quad \mathbb{P} \otimes \lambda|_{[0,t]}\text{-almost everywhere,}$$
$$F^k(s,u) \to 0, \quad \mathbb{P} \otimes \lambda|_{[0,t]} \otimes \beta\text{-almost everywhere.}$$

Furthermore, we define $Y, Y^k : \Omega \times [0,t] \to \mathbb{R}$ and $G, G^k : \Omega \times [0,t] \times E \to \mathbb{R}$ as

$$Y_s = 2c_4 K(\gamma + 1)e^{ct}(1 + \|Z_s^x\|^2),$$
$$Y_s^k = 2c_4 K(\gamma + 1)e^{ct}(1 + \|Z_s^{n_k,x}\|^2),$$
$$G(s,u) = 2c_5(\gamma + 1)e^{ct}C(1 + \|Z_s^x\|^2),$$
$$G^k(s,u) = 2c_5(\gamma + 1)e^{ct}C(1 + \|Z_s^{n_k,x}\|^2),$$

where the constants $K > 0$ and $C > 0$ stem from the linear growth conditions (7.1.3), (7.1.4), and the constant $\gamma > 0$ comes from Lemma A.3.1.

Using (7.1.11), Lemma A.3.1 and the linear growth condition (7.1.4) we obtain for all $k \in \mathbb{N}$ the estimate

$$|X_s^k| \leq 2c_4 K(\gamma + 1)e^{ct}(1 + \|Z_s^{n_k,x}\|^2) = Y_s^k$$

almost everywhere with respect to $\mathbb{P} \otimes \lambda|_{[0,t]}$. By Taylor's theorem, relation (7.1.12), Lemma A.3.1 and the linear growth condition (7.1.3) we get for all $k \in \mathbb{N}$ the estimate

$$|F^k(s,u)| \leq c_5 e^{c_2 t} \|n_k R_{n_k}(A)f(u, Z_s^{n_k,x})\|^2 + c_5 e^{c_2 t} \|f(u, Z_s^{n_k,x})\|^2$$
$$\leq 2c_5(\gamma + 1)e^{c_2 t}C(1 + \|Z_s^{n_k,x}\|^2) = G^k(s,u)$$

almost everywhere with respect to $\mathbb{P} \otimes \lambda|_{[0,t]} \otimes \beta$. This shows (7.1.13). By (7.1.17) we have

$$Y_s^k \to Y_s, \quad \mathbb{P} \otimes \lambda|_{[0,t]}\text{-almost everywhere},$$
$$G^k(s,u) \to G(s,u), \quad \mathbb{P} \otimes \lambda|_{[0,t]} \otimes \beta\text{-almost everywhere},$$

and, furthermore, by (7.1.16) we have

$$\mathbb{E}\left[\int_0^t Y_s^k ds\right] \to \mathbb{E}\left[\int_0^t Y_s ds\right],$$
$$\mathbb{E}\left[\int_0^t \int_E G^k(s,u)ds\beta(du)\right] \to \mathbb{E}\left[\int_0^t \int_E G(s,u)ds\beta(du)\right].$$

The generalized Lebesgue dominated convergence theorem (Lemma 7.1.8) applies and proves (7.1.15). $\qquad\square$

Theorem 7.1.10 *Suppose Assumption 7.1.1 is fulfilled, and there exists a Lyapunov function* $\psi \in C_b^{2,\mathrm{loc}}(H; \mathbb{R})$ *for* (7.0.1) *such that* (7.1.11) *and* (7.1.12) *are satisfied with appropriate constants* $c_4, c_5 > 0$. *Then, the SPDE* (7.0.1) *is exponentially stable in the mean square sense.*

Proof Let $x \in \mathcal{D}(A)$ and $t \geq 0$ be arbitrary. By Lemma 7.1.9, there exists a subsequence $(n_k)_{k \in \mathbb{N}}$ such that (7.1.13)–(7.1.15) are satisfied. Using (7.1.7) and (7.1.13), for each $k \in \mathbb{N}$ we have

$$\mathbb{E}\left[\int_0^t |e^{c_2 s}(c_2 + \mathcal{L}_{n_k})\psi(Z_s^{n_k,x})|ds\right] \leq \mathbb{E}\left[\int_0^t |e^{c_2 s}(\mathcal{L}_{n_k} - \mathcal{L})\psi(Z_s^{n_k,x})|ds\right] < \infty.$$

By Lemma 7.1.7 and relation (7.1.7) we obtain

$$e^{c_2 t}\mathbb{E}[\psi(Z_t^{n_k,x})] - \psi(x) = \mathbb{E}\left[\int_0^t e^{c_2 s}(c_2 + \mathcal{L}_{n_k})\psi(Z_s^{n_k,x})ds\right]$$
$$\leq \mathbb{E}\left[\int_0^t e^{c_2 s}(\mathcal{L}_{n_k} - \mathcal{L})\psi(Z_s^{n_k,x})ds\right], \quad k \in \mathbb{N}$$

and hence, by virtue of (7.1.14) and (7.1.15), we get

$$e^{c_2 t}\mathbb{E}[\psi(Z_t^x)] \leq \psi(x).$$

Incorporating (7.1.6) we obtain

$$c_1\mathbb{E}[\|Z_t^x\|^2] \leq \mathbb{E}[\psi(Z_t^x)] \leq e^{-c_2 t}\psi(x) \leq c_3 e^{-c_2 t}\|x\|^2.$$

Since $\mathcal{D}(A)$ is dense in H, applying Corollary 5.3.2 yields (7.1.5) with $c = \frac{c_3}{c_1}$ and $\theta = c_2$. $\qquad\qquad\qquad\qquad\qquad\qquad\qquad\qquad\qquad\qquad\qquad\qquad\qquad\qquad\square$

Now we want to show that in the linear case, if the SPDE has a zero solution (i.e. $x = 0$) and it is exponentially stable in the mean square sense, then we can construct a Lyapunov function. Let $f_0 : E \to \mathbb{R}$ be a function. We consider the linear SPDE

$$\begin{cases} dZ_t^0 = AZ_t^0 dt + \int_E f_0(v)Z_t^0 q(dt, dv), \\ Z_0^0 = x. \end{cases} \qquad (7.1.18)$$

Assumption 7.1.11 We assume that $d := \int_E f_0(y)^2 \beta(dy) < \infty$.

By Theorem 5.3.1, for each $x \in H$ there exists a unique mild solution $Z^{0,x}$ for (7.1.18) with $Z_0^0 = x$. For $n \in \mathbb{N}$ with $n > \alpha$ we consider the approximative system

$$\begin{cases} dZ_t^n = A_n Z_t^n dt + \int_E f_0(v)Z_t^n q(dt, dv), \\ Z_0^n = x, \end{cases} \qquad (7.1.19)$$

where $A_n \in L(H)$ denotes the Yosida approximation defined in Chap. 5. By Theorem 4.2.2, for each $x \in H$ there exists a unique strong solution $Z^{n,x}$ for (7.1.19) with $Z_0^{n,x} = x$.

We denote by $\mathcal{L}_0$ the infinitesimal generator for (7.1.18), and by $\mathcal{L}_n$ we denote the infinitesimal generator for (7.1.19).

Lemma 7.1.12 *Let $T \in L(H)$ be a self-adjoint operator. Then the function*

$$\psi : H \to \mathbb{R}, \quad \psi(x) = \langle Tx, x \rangle \qquad (7.1.20)$$

belongs to $C_b^{2,\mathrm{loc}}(H; \mathbb{R})$, there are constants $c_3, c_4, c_5 > 0$ such that (7.1.8), (7.1.11), (7.1.12) are satisfied, and we have

$$\mathcal{L}\psi(x) = 2\langle Tx, Ax + a(x)\rangle + \int_E \langle Tf(u, x), f(u, x)\rangle \beta(du), \quad x \in \mathcal{D}(A).$$

Proof This is a direct calculation.   □

Corollary 7.1.13 *Let* $T \in L(H)$ *be a self-adjoint operator. Then the function* (7.1.20) *belongs to* $C_b^{2,\mathrm{loc}}(H; \mathbb{R})$, *and we have*

$$\mathcal{L}_0\psi(x) = 2\langle Tx, Ax\rangle + d\langle Tx, x\rangle, \quad x \in \mathcal{D}(A)$$
$$\mathcal{L}_n\psi(x) = 2\langle Tx, A_n x\rangle + d\langle Tx, x\rangle, \quad x \in H$$

for all $n \in \mathbb{N}$.

Proof This is a direct consequence of Lemma 7.1.12.   □

Lemma 7.1.14 *Let* $B : H \times H \to \mathbb{R}$ *be a symmetric, bilinear operator. Then we have*

$$B(x, y) = \frac{1}{4}\big(B(x + y, x + y) - B(x - y, x - y)\big), \quad x, y \in H.$$

Proof This is a straightforward calculation by using the symmetry of the bilinear operator B.   □

Lemma 7.1.15 *Let* $B : H \times H \to \mathbb{R}$ *be a symmetric, bilinear operator and let* $B_n : H \times H \to \mathbb{R}$, $n \in \mathbb{N}$ *be a sequence of symmetric, bilinear operators such that* $B_n(x, x) \to B(x, x)$ *for all* $x \in H$. *Then we have*

$$B_n(x, y) \to B(x, y), \quad x, y \in H.$$

Proof This is a direct consequence of Lemma 7.1.14.   □

Lemma 7.1.16 *Let* $B : H \times H \to \mathbb{R}$ *be a symmetric, bilinear operator. Assume there exists a constant* $M > 0$ *such that*

$$\|B(x, x)\| \leq M \quad \text{for all } x \in H \text{ with } \|x\| = 1.$$

Then the bilinear operator B is continuous.

Proof By assumption we have

$$\|B(x, x)\| \leq \|x\|^2 B\left(\frac{x}{\|x\|}, \frac{x}{\|x\|}\right) \leq M\|x\|^2, \quad x \in H.$$

By Lemma 7.1.14 and the parallelogram identity we obtain

$$4\|B(x, y)\| \leq M(\|x + y\|^2 + \|x - y\|^2) = 2M(\|x\|^2 + \|y\|^2), \quad x, y \in H$$

proving the continuity of B.   □

Lemma 7.1.17 *Let $t \geq 0$ be arbitrary. There exist positive semidefinite, self-adjoint operators $(T_t^n)_{n \in \mathbb{N}_0} \subset L(H)$ such that*

$$\langle T_t^n x, y \rangle = \int_0^t \mathbb{E}[\langle Z_s^{n,x}, Z_s^{n,y} \rangle] ds, \quad n \in \mathbb{N}_0 \qquad (7.1.21)$$

for all $x, y \in H$, and we have

$$\lim_{n \to \infty} \mathcal{L}_n \psi_t^n(x) = \mathcal{L}_0 \psi_t^0(x), \quad x \in \mathcal{D}(A) \qquad (7.1.22)$$

where we have defined the functions $(\psi_t^n)_{n \in \mathbb{N}_0} \subset C_b^{2,loc}(H; \mathbb{R})$ by $\psi_t^n(x) = \langle T_t^n x, x \rangle$ for $n \in \mathbb{N}_0$.

Proof Let $n \in \mathbb{N}_0$ be arbitrary. Let $B_n : H \times H \to \mathbb{R}$ be the symmetric, bilinear operator

$$B_n(x, y) := \int_0^t \mathbb{E}[\langle Z_s^{n,x}, Z_s^{n,y} \rangle] ds.$$

Since $Z^{n,0} \equiv 0$, and the solution map $H \to S_t^2(H), x \mapsto Z^{n,x}$ is Lipschitz continuous by Corollary 5.3.2, we obtain

$$B_n(x, x) = \int_0^t \mathbb{E}[\|Z_s^{n,x}\|^2] \leq t \mathbb{E}\left[\sup_{s \in [0,t]} \|Z_s^{n,x}\|^2 \right] \leq tM\|x\|^2, \quad x \in H$$

for a suitable constant $M > 0$. By Lemma 7.1.16, the bilinear operator B_n is continuous. Thus, there exist positive semidefinite, self-adjoint operators $(T_t^n)_{n \in \mathbb{N}_0} \subset L(H)$ such that (7.1.21) is satisfied for all $x, y \in H$. By Theorem 5.3.3 we have

$$\lim_{n \to \infty} B_n(x, x) = \lim_{n \to \infty} \int_0^t \mathbb{E}[\|Z_s^{n,x}\|^2] ds = \int_0^t \mathbb{E}[\|Z_s^{0,x}\|^2] ds = B(x, x), \quad x \in H.$$

Lemma 7.1.15 implies that $\langle T_t^n x, y \rangle \to \langle T_t^0 x, y \rangle$ for all $x, y \in H$. Therefore, Corollary 7.1.13 and Theorem 5.3.3 yield the claimed convergence (7.1.22). $\square$

Lemma 7.1.18 *For all $t \geq 0$ and $x \in \mathcal{D}(A)$ we have*

$$\mathbb{E}[\|Z_t^{0,x}\|^2] = \mathcal{L}_0 \psi_t(x) + \|x\|^2.$$

Proof Let $t \geq 0$ and $x \in \mathcal{D}(A)$ be arbitrary. Furthermore, let $(P_t)_{t \geq 0} = (P_{0,t})_{t \geq 0}$ be the Markov semigroup in (7.1.18) as defined earlier in Sect. 5.4.

Let $s \in [0, t]$ and $n \in \mathbb{N}$ be arbitrary. Setting $\varphi : H \to \mathbb{R}, \varphi(h) = \|h\|^2$ we obtain, by using the Markov property,

$$\mathbb{E}[\psi_t^n(Z_s^{n,x})] = \mathbb{E}\left[\int_0^t \mathbb{E}[\varphi(Z_t^y)]|_{y=Z_s^{n,x}}ds\right] = \mathbb{E}\left[\int_0^t (P_u\varphi)(Z_s^{n,x})du\right]$$

$$= \mathbb{E}\left[\int_0^t \mathbb{E}[\varphi(Z_{u+s}^{n,x}) \mid \mathcal{F}_s^{Z^{n,x}}]du\right]$$

$$= \int_0^t \mathbb{E}[\|Z_{u+s}^{n,x}\|^2]du = \psi_{t+s}^n(x) - \psi_s^n(x), \qquad (7.1.23)$$

and Lemma 7.1.7 gives us

$$\mathbb{E}[\psi_t^n(Z_s^{n,x})] = \psi_t^n(x) + \int_0^s \mathbb{E}[(\mathcal{L}_n\psi_t^n)(Z_u^{n,x})]du. \qquad (7.1.24)$$

Combining (7.1.23) and (7.1.24) we get

$$\psi_{t+s}^n(x) - \psi_s^n(x) = \int_0^s \mathbb{E}[(\mathcal{L}_n\psi_t^n)(Z_u^{n,x})]du + \psi_t^n(x). \qquad (7.1.25)$$

Note that, moreover, we have

$$\lim_{s\to 0} \frac{\psi_s^n(x)}{s} = \frac{d}{ds}\psi_s^n(x)|_{s=0} = \mathbb{E}[\|Z_0^{n,x}\|^2] = \|x\|^2. \qquad (7.1.26)$$

Combining (7.1.25) and (7.1.26) we obtain

$$\frac{d}{dt}\psi_t^n(x) = \lim_{s\to 0} \frac{\psi_{t+s}^n(x) - \psi_t^n(x)}{s} = \lim_{s\to 0} \frac{\psi_s^n(x)}{s} + \mathbb{E}[\mathcal{L}_n\psi_t^n(Z_0^{n,x})]$$
$$= \|x\|^2 + \mathcal{L}_n\psi_t^n(x). \qquad (7.1.27)$$

Now, Theorem 5.3.3 and (7.1.27) yield

$$\frac{d}{dt}\psi_t(x) = \mathbb{E}[\|Z_t^{0,x}\|^2] = \lim_{n\to\infty} \mathbb{E}[\|Z_t^{n,x}\|^2] = \lim_{n\to\infty} \frac{d}{dt}\psi_t^n(x) = \mathcal{L}_0\psi_t(x) + \|x\|^2,$$
$$(7.1.28)$$

completing the proof.   □

Lemma 7.1.19 *Suppose the SPDE (7.1.18) is exponentially stable in the mean square sense. Then there exists a positive semidefinite, self-adjoint operator $T \in L(H)$ such that*

$$\langle Tx, y \rangle = \int_0^\infty \mathbb{E}[\langle Z_s^{0,x}, Z_s^{0,y}\rangle]ds, \quad x, y \in H \qquad (7.1.29)$$

and we have $\|T\| \le \frac{c}{\theta}$, where the constants $c, \theta > 0$ stem from (7.1.5).

Proof Let $B : H \times H \to \mathbb{R}$ be the symmetric, bilinear operator

$$B(x, y) := \int_0^\infty \mathbb{E}[\langle Z_s^x, Z_s^y \rangle] ds.$$

Using the estimate (7.1.5), we obtain

$$\|B(x, x)\| = \int_0^\infty \mathbb{E}[\|Z_s^{0,x}\|^2] ds \le c\|x\|^2 \int_0^\infty e^{-\theta s} ds = \frac{c}{\theta}\|x\|^2. \qquad (7.1.30)$$

By Lemma 7.1.16, the bilinear operator B is continuous. Hence, there exists a positive semidefinite, self-adjoint operator $T \in L(H)$ such that (7.1.29) is valid. Since $\|T\| = \sup_{\|x\| \le 1} |\langle Tx, x \rangle|$, estimate (7.1.30) shows that $\|T\| \le \frac{c}{\theta}$. $\qquad \square$

Since $(S_t)_{t \ge 0}$ is pseudo-contractive, there exists, by the Lumer–Phillips theorem, a constant $\lambda \ge 0$ such that

$$\langle Ax, x \rangle \le \lambda\|x\|^2 \quad \text{for all } x \in \mathcal{D}(A). \qquad (7.1.31)$$

Theorem 7.1.20 *Suppose Assumption 7.1.11 is fulfilled. If the linear SPDE (7.1.18) is exponentially stable in the mean square sense, then for each $\omega \in (0, \frac{1}{2\lambda+d})$ the function*

$$\Lambda_\omega^0 : H \to \mathbb{R}, \quad \Lambda_\omega^0(x) = \int_0^\infty \mathbb{E}[\|Z_s^{0,x}\|^2] ds + \omega\|x\|^2 \qquad (7.1.32)$$

is a Lyapunov function for (7.1.18), conditions (7.1.11) and (7.1.12) are satisfied with $\psi = \Lambda_\omega^0$ for suitable constants $c_4, c_5 > 0$, and we have the estimate

$$\mathcal{L}_0 \Lambda_\omega^0(x) \le -(1 - (2\lambda + d)\omega)\|x\|^2, \quad x \in \mathcal{D}(A). \qquad (7.1.33)$$

Proof Relation (7.1.5) with $Z = Z^0$ shows that for all $x \in H$ we obtain

$$\mathbb{E}[\|Z_t^{0,x}\|^2] \le ce^{-\theta t}\|x\|^2 \to 0 \quad \text{as } t \to \infty. \qquad (7.1.34)$$

Moreover, we have

$$\lim_{t \to \infty} \langle T_t x, x \rangle = \lim_{t \to \infty} \int_0^t \mathbb{E}[\|Z_s^x\|^2] ds = \int_0^\infty \mathbb{E}[\|Z_s^x\|^2] ds = \langle Tx, x \rangle, \quad x \in H.$$

Lemma 7.1.15 implies that $\langle T_t x, y \rangle \to \langle Tx, y \rangle$ for all $x, y \in H$. Therefore, Corollary 7.1.13, Lemma 7.1.18 and (7.1.34) yield

$$\mathcal{L}_0 \psi(x) = \lim_{t \to \infty} \mathcal{L}_0 \psi_t(x) = \lim_{t \to \infty} (\mathbb{E}[\|Z_t^{0,x}\|^2] - \|x\|^2) = -\|x\|^2, \quad x \in \mathcal{D}(A).$$

$$(7.1.35)$$

Let $\omega \in (0, \frac{1}{2\lambda+d})$ be arbitrary. The function $\Lambda_\omega^0 : H \to \mathbb{R}$ defined in (7.1.32) has the representation

$$\Lambda_\omega^0(x) = \langle (T + \omega)x, x \rangle = \langle Tx, x \rangle + \omega \|x\|^2, \quad x \in H.$$

Thus, by Lemma 7.1.12, the function Λ_ω^0 belongs to $C_b^{2,\text{loc}}(H; \mathbb{R})$, and there are constants $c_4, c_5 > 0$ such that (7.1.11) and (7.1.12) are satisfied with $\psi = \Lambda_\omega^0$. Taking into account Lemma 7.1.19, we have

$$\omega \|x\|^2 \leq \Lambda_0(x) \leq \left(\frac{c}{\theta} + \omega \right) \|x\|^2, \quad x \in H \tag{7.1.36}$$

that is, condition (7.1.6) is satisfied with $c_1 = \omega$, $c_3 = \frac{c}{\theta} + \omega$ and $\psi = \Lambda_\omega^0$. By Lemma 7.1.13, relations (7.1.35), (7.1.31) and (7.1.36) we obtain

$$
\begin{aligned}
\mathcal{L}_0 \Lambda_\omega^0(x) &= 2\langle (T + \omega)x, Ax \rangle + d \langle (T + \omega)x, x \rangle \\
&= \mathcal{L}_0 \psi(x) + (2\langle Ax, x \rangle + d\|x\|^2)\omega \\
&\leq -\|x\|^2 + \omega(2\lambda + d)\|x\|^2 = -(1 - (2\lambda + d)\omega)\|x\|^2 \\
&\leq -\left(\frac{1 - (2\lambda + d)\omega}{\frac{c}{\theta} + \omega} \right) \Lambda_\omega^0(x), \quad x \in \mathcal{D}(A),
\end{aligned}
$$

proving the estimate (7.1.33) and condition (7.1.7) is satisfied with $c_2 = \frac{1 - (2\lambda+d)\omega}{\frac{c}{\theta} + \omega}$, $\psi = \Lambda_\omega^0$ and $\mathcal{L} = \mathcal{L}_0$. $\qquad\qquad\qquad\qquad\qquad\qquad\qquad\qquad\qquad\qquad\qquad\qquad\square$

The reason for looking at the linear case is that, in general, we do not know if

$$\psi(x) = \mathbb{E}\left[\int_0^\infty \|Z_t^x\|^2 dt \right]$$

is in $C_b^{2,\text{loc}}(H; \mathbb{R})$. However, using the results of Sect. 5.7, we can give conditions on the coefficient for $\psi \in C_b^{2,\text{loc}}(H; \mathbb{R})$. In this case

$$\Lambda(x) = \psi(x) + \omega \|x\|^2$$

satisfies condition (7.1.6).

Lemma 7.1.21 *Let $T \in L(H)$ be a self-adjoint operator. Then we have*

$$|\langle Tx, x \rangle - \langle Ty, y \rangle| \leq \|T\| \, \|x - y\| \, \|x + y\|, \quad x, y \in H.$$

Proof Let $x, y \in H$ be arbitrary. Then we have

$$\langle Tx, x \rangle - \langle Ty, y \rangle = \langle Tx, x \rangle + \langle Tx, y \rangle - \langle Ty, x \rangle - \langle Ty, y \rangle$$
$$= \langle Tx, x + y \rangle - \langle Ty, x + y \rangle = \langle T(x - y), x + y \rangle,$$

which yields the assertion.   □

Theorem 7.1.22 *Suppose Assumptions* 7.1.1 *and* 7.1.11 *are fulfilled, the solution of the linear SPDE* (7.1.18) *is exponentially stable in the mean square sense, and there is a constant* $\epsilon > 0$ *such that for all* $x \in H$ *we have*

$$2\|x\|\|a(x)\| + \int_E \|f(v, x) - f_0(v)x\|\|f(v, x) + f_0(v)x\|\beta(dv) \leq (1 - \epsilon)\frac{\theta}{c}\|x\|^2,$$
(7.1.37)

where the constants $c, \theta > 0$ *stem from* (7.1.5). *Then the solution of the SPDE* (7.0.1) *is also exponentially stable in the mean square sense.*

Proof There exists an $\omega \in (0, \frac{1}{2\lambda + d})$ such that

$$C_\omega := \epsilon - \left(2\lambda + d + (1 - \epsilon)\frac{\theta}{c}\right)\omega > 0.$$
(7.1.38)

By Theorem 7.1.20, the function $\Lambda_\omega^0 : H \to \mathbb{R}$ defined in (7.1.32) is a Lyapunov function for (7.1.18), conditions (7.1.11) and (7.1.12) are satisfied with $\psi = \Lambda_\omega^0$ for suitable constants $c_4, c_5 > 0$, and we have the estimate (7.1.33). Note that we have the representation

$$\Lambda_\omega^0(x) = \langle Tx, x \rangle + \omega\|x\|^2 = \langle (T + \omega)x, x \rangle, \quad x \in H.$$

Let $x \in \mathcal{D}(A)$ be arbitrary. Using Lemma 7.1.12, estimate (7.1.37) and Lemma 7.1.21 we obtain

$$\mathcal{L}\Lambda_\omega^0(x) - \mathcal{L}_0\Lambda_\omega^0(x) = 2\langle (T + \omega)x, a(x) \rangle$$
$$+ \int_E \Big(\langle (T + \omega)f(v, x), f(v, x) \rangle$$
$$- \langle (T + \omega)f_0(v)x, f_0(v)x \rangle \Big)\nu(dv)$$
$$\leq (\|T\| + \omega)\bigg(2\|x\|\|a(x)\| + \int_{H \setminus \{0\}} \|f(v, x)$$
$$- f_0(v)x\|\|f(v, x) + f_0(v)x\|\nu(dv)\bigg)$$
$$\leq (1 - \epsilon)\left(\frac{c}{\theta} + \omega\right)\frac{\theta}{c}\|x\|^2 = (1 - \epsilon)\left(1 + \frac{\omega\theta}{c}\right)\|x\|^2,$$

and therefore, by taking into account (7.1.33),

$$\mathcal{L}\Lambda^0_\omega(x) \leq \mathcal{L}_0\Lambda^0_\omega(x) + (1 - \epsilon)\left(1 + \frac{\omega\theta}{c}\right)\|x\|^2$$

$$\leq -(1 - (2\lambda + d)\omega)\|x\|^2 + (1 - \epsilon)\left(1 + \frac{\omega\theta}{c}\right)\|x\|^2 = -C_\omega\|x\|^2.$$

By (7.1.38) and (7.1.6), condition (7.1.7) is satisfied with $\psi = \Lambda^0_\omega$ and a suitable constant $c_2 > 0$. Consequently, Λ^0_ω is also a Lyapunov function for (7.0.1). By Theorem 7.1.10, the SPDE (7.0.1) is exponentially stable in the mean square sense.   □

Definition 7.1.23 We say that the zero solution of (7.0.1) is stable in probability if for each $\epsilon > 0$ we have

$$\lim_{\|x\|\to 0} \mathbb{P}\left(\sup_{t\geq 0} \|Z_t^x\| > \epsilon\right) = 0. \tag{7.1.39}$$

Theorem 7.1.24 *Suppose Assumption 7.1.1 is fulfilled, and there exists a function $\psi \in C_b^{2,\mathrm{loc}}(H;\mathbb{R})$ and constants $c_1, c_3 > 0$ such that (7.1.6) and (7.1.7) are satisfied with $c_2 = 0$, and (7.1.11) and (7.1.12) are satisfied with appropriate constants $c_4, c_5 > 0$. Then the zero solution of (7.0.1) is stable in probability.*

Proof Let $x \in \mathcal{D}(A)$ and $t \geq 0$ be arbitrary. By Lemma 7.1.9, there exists a subsequence $(n_k)_{k\in\mathbb{N}}$ such that (7.1.13)–(7.1.15) are satisfied with $c_2 = 0$. Using (7.1.7) and (7.1.13), for each $k \in \mathbb{N}$ we have

$$\mathbb{E}\left[\int_0^t |\mathcal{L}_{n_k}\psi(Z_s^{n_k,x})|ds\right] \leq \mathbb{E}\left[\int_0^t |(\mathcal{L}_{n_k} - \mathcal{L})\psi(Z_s^{n_k,x})|ds\right] < \infty.$$

By Lemma 7.1.7 and relation (7.1.7) we obtain

$$\mathbb{E}[\psi(Z_t^{n_k,x})] - \psi(x) = \mathbb{E}\left[\int_0^t \mathcal{L}_{n_k}\psi(Z_s^{n_k,x})ds\right]$$

$$\leq \mathbb{E}\left[\int_0^t (\mathcal{L}_{n_k} - \mathcal{L})\psi(Z_s^{n_k,x})ds\right], \quad k \in \mathbb{N}$$

and hence, by virtue of (7.1.14) and (7.1.15), we get

$$\mathbb{E}[\psi(Z_t^x)] \leq \psi(x).$$

Now, let $x \in H$ be arbitrary. Since $\mathcal{D}(A)$ is dense in H, there exists a sequence $(x_n)_{n\in\mathbb{N}} \subset \mathcal{D}(A)$ with $x_n \to x$. By (7.1.8), Corollary 5.3.2 and the generalized Lebesgue dominated convergence theorem (Lemma 7.1.8), there exists a subsequence $(n_k)_{k\in\mathbb{N}}$ such that

$$\mathbb{E}[\psi(Z_t^{x_{n_k}})] \rightarrow \mathbb{E}[\psi(Z_t^x)].$$

Hence, we deduce that

$$\mathbb{E}[\psi(Z_t^x)] \leq \psi(x) \quad \text{for all } x \in H \text{ and } t \geq 0. \tag{7.1.40}$$

Now, let $x \in H$ and $\epsilon > 0$ be arbitrary. We define the stopping time

$$\tau_\epsilon^x := \inf\{t \geq 0 : \|Z_t^x\| > \epsilon\}.$$

By (7.1.6) and (7.1.40) we obtain

$$c_1 \epsilon \mathbb{P}(\tau_\epsilon^x \leq t) \leq \mathbb{E}[\psi(Z_{t \wedge \tau_\epsilon^x}^x)] \leq \psi(x),$$

which implies

$$\mathbb{P}\left(\sup_{t \geq 0} \|Z_t^x\| > \epsilon\right) \leq \mathbb{P}(\tau_\epsilon^x \leq t) \leq \frac{\psi(x)}{c_1 \epsilon},$$

and hence, in view of (7.1.6), we arrive at (7.1.39). □

Corollary 7.1.25 *Suppose Assumption 7.1.11 is fulfilled. If the linear SPDE (7.1.18) is exponentially stable in the mean square sense, then the zero solution of (7.1.18) is stable in probability.*

Proof This is a direct consequence of Theorems 7.1.20, 7.1.10 and 7.1.24. □

Corollary 7.1.26 *Suppose Assumption 7.1.1 and 7.1.11 are fulfilled. If the linear SPDE (7.1.18) is exponentially stable in the mean square sense and we have (7.1.37), then the zero solution of (7.0.1) is stable in probability.*

Proof This is an immediate consequence of Theorems 7.1.22, 7.1.10 and 7.1.24. □

7.2 Exponential Ultimate Boundedness in the Mean Square Sense

Definition 7.2.1 The solution of SPDE (7.0.1) is called exponentially ultimately bounded in the mean square sense if there exist constants $c, \theta, M > 0$ such that

$$\mathbb{E}[\|Z_t^x\|^2] \leq ce^{-\theta t}\|x\|^2 + M \quad \text{for all } x \in H \text{ and } t \geq 0. \tag{7.2.1}$$

Theorem 7.2.2 *Suppose Assumption 7.1.1 is fulfilled, and there exist a function $\psi \in C_b^{2,\text{loc}}(H; \mathbb{R})$ and constants $c_i > 0$, $i = 1, \ldots, 5$ and $k_i \geq 0$, $i = 1, 2, 3$ such that (7.1.11) and (7.1.12) are satisfied and we have*

$$c_1 \|x\|^2 - k_1 \leq \psi(x) \leq c_3 \|x\|^2 - k_3, \quad x \in H, \tag{7.2.2}$$

$$\mathcal{L}\psi(x) \leq -c_2 \psi(x) + k_2, \quad x \in \mathcal{D}(A). \tag{7.2.3}$$

Then the solution of SPDE (7.0.1) is exponentially ultimately bounded.

Proof Let $x \in \mathcal{D}(A)$ and $t \geq 0$ be arbitrary. By Lemma 7.1.9, there exists a subsequence $(n_k)_{k \in \mathbb{N}}$ such that (7.1.13)–(7.1.15) are satisfied. Using (7.1.7) and (7.1.13), for each $k \in \mathbb{N}$ we have

$$\mathbb{E}\left[\int_0^t |e^{c_2 s}(c_2 + \mathcal{L}_{n_k})\psi(Z_s^{n_k,x})| ds\right] \leq \mathbb{E}\left[\int_0^t |e^{c_2 s}((\mathcal{L}_{n_k} - \mathcal{L})\psi(Z_s^{n_k,x}) + k_2)| ds\right]$$

$$\leq \mathbb{E}\left[\int_0^t |e^{c_2 s}(\mathcal{L}_{n_k} - \mathcal{L})\psi(Z_s^{n_k,x})| ds\right]$$

$$+ k_2 \int_0^t e^{c_2 s} ds < \infty.$$

By Lemma 7.1.7 and relation (7.1.7) we obtain

$$e^{c_2 t}\mathbb{E}[\psi(Z_t^{n_k,x})] - \psi(x) = \mathbb{E}\left[\int_0^t e^{c_2 s}(c_2 + \mathcal{L}_{n_k})\psi(Z_s^{n_k,x}) ds\right]$$

$$\leq \mathbb{E}\left[\int_0^t e^{c_2 s}((\mathcal{L}_{n_k} - \mathcal{L})\psi(Z_s^{n_k,x}) + c_2) ds\right]$$

$$= \mathbb{E}\left[\int_0^t e^{c_2 s}(\mathcal{L}_{n_k} - \mathcal{L})\psi(Z_s^{n_k,x}) ds\right] + k_2 \int_0^t e^{c_2 s} ds,$$

and hence, by virtue of (7.1.14) and (7.1.15), we get

$$e^{c_2 t}\mathbb{E}[\psi(Z_t^x)] \leq \psi(x) + \frac{k_2}{c_2}(e^{c_2 t} - 1).$$

Incorporating (7.2.2) we obtain

$$c_1 \mathbb{E}[\|Z_t^x\|^2] - k_1 \leq \mathbb{E}[\psi(Z_t^x)] \leq e^{-c_2 t}\left(\psi(x) + \frac{k_2}{c_2}(e^{c_2 t} - 1)\right)$$

$$\leq e^{-c_2 t}(c_3 \|x\|^2 - k_3) + \frac{k_2}{c_2}(1 - e^{-c_2 t}) \leq c_3 e^{-c_2 t}\|x\|^2 + \frac{k_2}{c_2}.$$

Since $\mathcal{D}(A)$ is dense in H, applying Corollary 5.3.2 yields (7.2.1) with

$$c = \frac{c_3}{c_1}, \quad \theta = c_2 \quad \text{and} \quad M = \frac{1}{c_1}\left(k_1 + \frac{k_2}{c_2}\right).$$

This completes the proof. $\square$

Corollary 7.2.3 *Suppose the assumptions of Theorem 7.2.2 are fulfilled. Then, for all $x \in H$, there exists a finite constant $M > 0$ such that*

$$\limsup_{t \to \infty} \mathbb{E}[\|Z_t^x\|^2] \leq M.$$

Proof The assertion follows from Theorem 7.2.2 and the estimate (7.2.1). □

Remark 7.2.4 The above Corollary 7.2.3 generalizes a result of Skorokhod ([97], p. 70).

Lemma 7.2.5 *Suppose the SPDE (7.1.18) is exponentially ultimately bounded in the mean square sense and let $t \geq 0$ be arbitrary. Then we have $\|T_t\| \leq \frac{c}{\theta} + Mt$, where $T_t \in L(H)$ denotes the positive semidefinite, self-adjoint operator from Lemma 7.1.17, and where the constants $c, \theta, M > 0$ stem from (7.2.1).*

Proof Using the estimate (7.2.1), for all $x \in H$ we obtain

$$\|\langle T_t x, x \rangle\| = \int_0^t \mathbb{E}[\|Z_s^{0,x}\|^2] ds \leq c\|x\|^2 \int_0^t e^{-\theta s} ds + Mt = \frac{c}{\theta}\|x\|^2 + Mt.$$

Since $\|T_t\| = \sup_{\|x\| \leq 1} |\langle T_t x, x \rangle|$, we deduce that $\|T_t\| \leq \frac{c}{\theta} + Mt$. □

Theorem 7.2.6 *If the solution of the linear SPDE (7.1.18) is exponentially ultimately bounded in the mean square sense, then for each $t > \frac{\ln c}{\theta}$ and each $\omega \in (0, \frac{ce^{\theta t}-1}{2\lambda+d})$ there are constants $c_i > 0$, $i = 1, \ldots, 5$ and $k_i \geq 0$, $i = 1, 2, 3$ such that the function*

$$\Lambda_{\omega,t}^0 : H \to \mathbb{R}, \quad \Lambda_{\omega,t}^0(x) = \int_0^t \mathbb{E}[\|Z_s^{0,x}\|^2] ds + \omega\|x\|^2 \qquad (7.2.4)$$

satisfies (7.1.11), (7.1.12) and (7.2.2), (7.2.3) with $\psi = \Lambda_{\omega,t}^0$ and $\mathcal{L} = \mathcal{L}_0$, and we have the estimate

$$\mathcal{L}_0 \Lambda_{\omega,t}^0(x) \leq -(1 - ce^{-\theta t} - (2\lambda + d)\omega)\|x\|^2 + M, \quad x \in \mathcal{D}(A) \qquad (7.2.5)$$

where the constants $c, \theta, M > 0$ stem from (7.2.1).

Proof By Lemma 7.1.18 and the estimate (7.2.1) we obtain

$$\mathcal{L}_0 \psi_t(x) = -\|x\|^2 + \mathbb{E}[\|Z_t^{0,x}\|^2] \leq -\|x\|^2 + ce^{-\theta t}\|x\|^2 + M$$
$$= (ce^{-\theta t} - 1)\|x\|^2 + M.$$

Let $t > \frac{\ln c}{\theta}$ and each $\omega \in (0, \frac{ce^{\theta t}-1}{2\lambda+d})$ be arbitrary. The function $\Lambda_{\omega,t}^0 : H \to \mathbb{R}$ defined in (7.2.4) has the representation

$$\Lambda^0_{\omega,t}(x) = \langle (T_t + \omega)x, x \rangle = \langle T_t x, x \rangle + \omega \|x\|^2, \quad x \in H.$$

Thus, by Lemma 7.1.12, the function $\Lambda^0_{\omega,t}$ belongs to $C^{2,\mathrm{loc}}_b(H;\mathbb{R})$, and there are constants $c_4, c_5 > 0$ such that (7.1.11) and (7.1.12) are satisfied with $\psi = \Lambda^0_{\omega,t}$. Using Lemma 7.2.5 we have

$$\omega\|x\|^2 \le \Lambda_0(x) \le \left(\frac{c}{\theta} + Mt + \omega\right)\|x\|^2, \quad x \in H, \tag{7.2.6}$$

that is, condition (7.2.2) is satisfied with $c_1 = \omega$, $c_3 = \frac{c}{\theta} + Mt + \omega$ and $\psi = \Lambda^0_\omega$. By Lemma 7.1.13, relations (7.1.35), (7.1.31) and (7.2.6) we obtain

$$
\begin{aligned}
\mathcal{L}_0\Lambda^0_{\omega,t}(x) &= 2\langle (T_t + \omega)x, Ax \rangle + d\langle (T_t + \omega)x, x \rangle \\
&= \mathcal{L}_0\psi_t(x) + (2\langle Ax, x \rangle + d\|x\|^2)\omega \\
&\le (ce^{-\theta t} - 1)\|x\|^2 + M + (2\lambda + d)\omega\|x\|^2 \\
&= -(1 - ce^{-\theta t} - (2\lambda + d)\omega)\|x\|^2 + M \\
&\le -\left(\frac{1 - ce^{-\theta t} - (2\lambda + d)\omega}{\frac{c}{\theta} + Mt + \omega}\right)\Lambda^0_{\omega,t}(x) + M, \quad x \in \mathcal{D}(A)
\end{aligned}
$$

proving the estimate (7.2.5) and condition (7.1.7) with

$$c_2 = \frac{1 - ce^{-\theta t} - (2\lambda + d)\omega}{\frac{c}{\theta} + Mt + \omega}$$

and $\mathcal{L} = \mathcal{L}_0$. $\square$

Theorem 7.2.7 *If the solution of equation (7.0.1) is exponentially ultimately bounded in the mean square sense and*

$$\phi(x) = \int_0^T \mathbb{E}[\|Z^x_s\|^2]ds$$

belongs to $C^{2,\mathrm{loc}}_b(H;\mathbb{R})$ for some $T > 0$, then there exists an $\omega \ge 0$ such that

$$\psi(x) = \phi(x) + \omega\|x\|^2$$

is a Lyapunov function.

Proof The proof is similar to that of Theorem 7.1.20. $\square$

Theorem 7.2.8 *Suppose Assumptions 7.1.1, 7.1.11 are fulfilled, the SPDE (7.1.18) is exponentially ultimately bounded in the mean square sense, and there exist constants $W, N > 0$ such that*

$$W < \max_{t > \frac{\ln c}{\theta}} \frac{1 - ce^{-\theta t}}{\frac{c}{\theta} + Mt} \qquad (7.2.7)$$

and for all $x \in H$ we have

$$2\|x\|\|a(x)\| + \int_{H \backslash \{0\}} \|f(v, x) - f_0(v)x\|\|f(v, x) + f_0(v)x\| \nu(dv) \le W\|x\|^2 + N, \qquad (7.2.8)$$

where the constants $c, \theta, M > 0$ stem from (7.2.1). Then the solution of the SPDE (7.0.1) is also exponentially ultimately bounded in the mean square sense.

Proof By (7.2.7), there exist $t > \frac{\ln c}{\theta}$ and $\epsilon > 0$ such that

$$W = (1 - \epsilon) \frac{1 - ce^{-\theta t}}{\frac{c}{\theta} + Mt}. \qquad (7.2.9)$$

Furthermore, there exists an $\omega \in (0, \frac{ce^{\theta t} - 1}{2\lambda + d})$ such that

$$C_{\omega, t} := (1 - ce^{-\theta t})\epsilon - (2\lambda + d + W)\omega > 0. \qquad (7.2.10)$$

By Theorem 7.1.20, there are constants $c_i > 0$, $i = 1, \ldots, 5$ and $k_i \ge 0$, $i = 1, 2, 3$ such that the function $\Lambda^0_\omega : H \to \mathbb{R}$ defined in (7.1.32) satisfies (7.1.11), (7.1.12) and (7.2.2), (7.2.3) with $\psi = \Lambda^0_{\omega, t}$ and $\mathcal{L} = \mathcal{L}_0$, and we have the estimate (7.2.5). Note that we have the representation

$$\Lambda^0_{t, \omega}(x) = \langle T_t x, x \rangle + \omega\|x\|^2 = \langle (T_t + \omega)x, x \rangle, \quad x \in H.$$

Let $x \in \mathcal{D}(A)$ be arbitrary. Using Lemmas 7.1.12, 7.1.21 and estimate (7.2.8) we obtain

$$\mathcal{L}\Lambda^0_{\omega, t}(x) - \mathcal{L}_0 \Lambda^0_{\omega, t}(x) = 2\langle (T_t + \omega)x, a(x) \rangle$$

$$+ \int_E \Big(\langle (T_t + \omega)f(v, x), f(v, x) \rangle$$

$$- \langle (T_t + \omega)f_0(v)x, f_0(v)x \rangle \Big)\beta(dv)$$

$$\le (\|T_t\| + \omega)\Big(2\|x\|\|a(x)\|$$

$$+ \int_{H \backslash \{0\}} \|f(v, x) - f_0(v)x\|\|f(v, x) + f_0(v)x\|\beta(dv) \Big)$$

$$\le \Big(\frac{c}{\theta} + Mt + \omega \Big)(W\|x\|^2 + N),$$

and therefore, by taking into account (7.2.5) and (7.2.9),

$$\mathcal{L}\Lambda^0_{\omega,t}(x) \leq -(1 - ce^{-\theta t} - (2\lambda + d)\omega)\|x\|^2 + M$$
$$+ \left(\frac{c}{\theta} + Mt + \omega\right)(W\|x\|^2 + N)$$
$$= -C_{t,\omega}\|x\|^2 + M + \left(\frac{c}{\theta} + Mt + \omega\right)N.$$

By (7.2.10) and (7.2.2), condition (7.2.3) is satisfied with $\psi = \Lambda^0_{\omega,t}$ and suitable constants $c_2 > 0$ and $k_2 \geq 0$. Consequently, $\Lambda^0_{\omega,t}$ is also a Lyapunov function for (7.0.1). By Theorem 7.2.6, the SPDE (7.0.1) is exponentially ultimately bounded in the mean square sense.   □

Let us recall some notation, which we will use in the sequel. For a function $f : H \to \mathbb{R}$ the notation

$$f(x) \to 0 \quad \text{for } \|x\| \to \infty$$

means that for each $\epsilon > 0$ there exists a constant $C > 0$ such that

$$\|f(x)\| \leq \epsilon \quad \text{for all } x \in H \text{ with } \|x\| \geq C.$$

For two functions $f, g : H \to \mathbb{R}$ the notation

$$f(x) = o(g(x)) \quad \text{for } \|x\| \to \infty$$

means that

$$\frac{f(x)}{g(x)} \to 0 \quad \text{for } \|x\| \to \infty.$$

Corollary 7.2.9 *Suppose the linear SPDE (7.1.18) is exponentially ultimately bounded in the mean square sense and*

$$\|a(x)\| = o(\|x\|), \qquad (7.2.11)$$

$$\int_E \|f(v, x) - f_0(v)x\| \|f(v, x) + f_0(v)x\| \nu(dv) = o(\|x\|^2) \qquad (7.2.12)$$

for $\|x\| \to \infty$. Then the solution of SPDE (7.0.1) is also exponentially ultimately bounded in the mean square sense.

Proof There exists a constant $W > 0$ such that (7.2.7) is satisfied. By (7.2.11) and (7.2.12) there exists a $C > 0$ such that

$$2\|x\|\|a(x)\| + \int_E \|f(v,x) - f_0(v)x\|\|f(v,x) + f_0(v)x\|v(dv) \le W\|x\|^2$$

for all $x \in H$ with $\|x\| \ge C$. Set

$$M := C^2 + 2K(C^2 + 1) + 2C^2,$$

where the constant $K > 0$ stems from the linear growth condition (7.1.4). For each $x \in H$ with $\|x\| < C$ we obtain, by using (7.1.4),

$$2\|x\|\|a(x)\| + \int_E \|f(v,x) - f_0(v)x\|\|f(v,x) + f_0(v)x\|\beta(dv)$$

$$\le \|x\|^2 + \|a(x)\|^2 + 2\int_E (\|f(v,x)\|^2\beta(dv) + 2d\|x\|^2$$

$$\le \|x\|^2 + 2K(\|x\|^2 + 1) + 2d\|x\|^2 < M.$$

Consequently, condition (7.2.8) is satisfied, whence Theorem 7.2.8 completes the proof. □

Corollary 7.2.10 *Suppose Assumption* 7.1.1 *is fulfilled, that the solution of the deterministic PDE*

$$\begin{cases} dZ_t = AZ_t dt, \\ Z_0 = x \end{cases}$$

is exponentially ultimately bounded and we have

$$\|a(x)\| = o(\|x\|),$$

$$\int_E \|f(v,x)\|^2 v(dv) = o(\|x\|^2)$$

for $\|x\| \to \infty$. *Then, the SPDE* (7.0.1) *is exponentially ultimately bounded in the mean square sense.*

Proof The assertion follows from Corollary 7.2.9 with $f_0 \equiv 0$. □

7.3 Invariant Measures

Let $(P_t)_{t\ge 0}$ be the Markov semigroup of (7.0.1) as defined in (5.4.2) for $s = 0$, i.e. P_t is the linear operator on $B_b(H)$ defined by

$$(P_t)(\phi)(x) = \mathbb{E}[\phi(Z_t^x)] \quad \text{for} \quad \phi \in B_b(H) \quad x \in H$$

where $Z_t^x := Z(t, 0; x)$ is the solution of (7.0.1) with initial condition $x \in H$ evaluated at time $t > 0$.

Definition 7.3.1 $(P_t)_{t \geq 0}$ is a Feller semigroup if $P_t(C_b(H)) \subset C_b(H)$ for all $t \in \mathbb{R}_+$, where $C_b(H)$ denotes the space of all bounded, continuous functions $f : H \to H$.

By continuous dependence on the initial condition (see Lemma 5.7.1) the semigroup $(P_t)_{t \geq 0}$ defined in (5.4.10) is a Feller semigroup.

Definition 7.3.2 [28, p. 230] A σ-finite measure μ on $(H, \mathcal{B}(H))$ is an invariant measure for $(P_t)_{t \geq 0}$ (resp. for the SPDE (7.0.1)) if we have

$$\int_H P_t f d\mu = \int_H f d\mu \quad \text{for all } f \in C_b(H) \text{ and } t \geq 0.$$

Definition 7.3.3 A sequence $(\mu_n)_{n \in \mathbb{N}}$ of probability measures on a separable metric space X *converges weakly* to a probability measure μ if

$$\int_X f d\mu_n \to \int_X f d\mu$$

for every $f \in C_b(X)$.

Let $(e_k)_{k \in \mathbb{N}}$ be an orthonormal basis of H, and define the isometric isomorphism

$$\mathcal{J} : H \to \ell^2(\mathbb{N}), \quad \mathcal{J}x := (\langle x, e_k \rangle)_{k \in \mathbb{N}}.$$

Note that $\ell^2(\mathbb{N}) \subset \mathbb{R}^\mathbb{N}$. The linear space $\mathbb{R}^\mathbb{N}$, equipped with the metric

$$d(x, y) = \sum_{k=1}^{\infty} 2^{-k} \frac{|x^k - y^k|}{1 + |x^k - y^k|}, \quad x, y \in \mathbb{R}^\mathbb{N}$$

is a separable metric space, and we have $x_n \to x$ in $\mathbb{R}^\mathbb{N}$ if and only if $x_n^k \to x^k$ for all $k \in \mathbb{N}$.

For a finite subset $I \subset \mathbb{N}$ we denote by $\pi_I : \mathbb{R}^\mathbb{N} \to \mathbb{R}^I$ the corresponding projection, and for finite subsets $I \subset J \subset \mathbb{N}$ we denote by $\pi_I^J : \mathbb{R}^J \to \mathbb{R}^I$ the corresponding projection. The family of $\ell^2(\mathbb{N})$-valued processes

$$Y_t^x := \mathcal{J} Z_t^{\mathcal{J}^{-1}(x)}, \quad x \in \ell^2(\mathbb{N}) \text{ and } t \geq 0$$

is a family of time-homogeneous Markov processes. We denote its Markov semigroup by $(Q_t)_{t \geq 0}$. For every finite subset $I \subset \mathbb{N}$, the family of $\mathbb{R}^I$-valued processes

$$Y_t^{I,x} := \pi_I Y_t^x, \quad x \in \mathbb{R}^I \text{ and } t \geq 0$$

is a family of time-homogeneous Markov processes. We denote its Markov semigroup by $(Q_t^I)_{t \geq 0}$.

Theorem 7.3.4 *Suppose Assumption 7.1.1 is fulfilled. If the solution of the SPDE (7.0.1) is exponentially ultimately bounded in the mean square sense, then it has an invariant measure ν satisfying*

$$\int_H \|x\|^2 \nu(dx) < \infty. \tag{7.3.1}$$

Proof Let $x \in H$ and a finite subset $I \subset \mathbb{N}$ be arbitrary. We show that the family of probability measures

$$\mu_I^{n,x}(B) := \frac{1}{n} \int_0^n \mathbb{P}(Y_s^{I,x} \in B) ds, \quad B \in \mathcal{B}(\mathbb{R}^I), \quad n \in \mathbb{N}$$

is tight. Indeed, for an arbitrary $\epsilon > 0$ we define the compact subset

$$K_\epsilon := \left\{ y \in \mathbb{R}^I : \|y\|_{\mathbb{R}^I} \leq \sqrt{\frac{c\|x\|^2 + M}{\epsilon}} \right\},$$

where the constants $c, M > 0$ stem from (7.2.1). Then, by Chebyshev's inequality and (7.2.1),

$$
\begin{aligned}
\mu_I^{n,x}(\mathbb{R}^I \setminus K_\epsilon) &= \frac{1}{n} \int_0^n \mathbb{P}\left(\|Y_s^{I,x}\|_{\mathbb{R}^I} > \sqrt{\frac{c\|x\|^2 + M}{\epsilon}} \right) ds \\
&\leq \frac{\epsilon}{n(c\|x\|^2 + M)} \int_0^n \mathbb{E}[\|Y_s^{I,x}\|_{\mathbb{R}^n}^2] ds \\
&\leq \frac{\epsilon}{n(c\|x\|^2 + M)} \int_0^n \mathbb{E}[\|Z_s^x\|^2] ds \leq \epsilon,
\end{aligned}
$$

proving the tightness. By Prokhorov's theorem, there exist a subsequence $(n_k)_{k \in \mathbb{N}}$ and a probability measure μ_I on $(\mathbb{R}^I, \mathcal{B}(\mathbb{R}^I))$ such that $\mu_I^{n_k,x} \to \mu_I$ weakly. According to [28, Theorem 9.3, p. 240], the probability measure μ_I is an invariant measure for the Markov semigroup $(Q_t^I)_{t \geq 0}$, that is, we have

$$\int_{\mathbb{R}^I} Q_t^I f_I d\mu_I = \int_{\mathbb{R}^I} f_I d\mu_I \quad \text{for all } f_I \in C_b(\mathbb{R}^I) \text{ and } t \geq 0. \tag{7.3.2}$$

The family $\{\mu_I : I \subset \mathbb{N} \text{ finite}\}$ is consistent, that is, for finite subsets $I \subset J \subset \mathbb{N}$ we have

$$\mu_I = \mu_J \circ \pi_I^J. \tag{7.3.3}$$

Indeed, let $I \subset J \subset \mathbb{N}$ be arbitrary finite subsets. Then for each $n \in \mathbb{N}$ we have $\mu_I^{n,x} = \mu_J^{n,x} \circ \pi_I^J$. There exists a joint subsequence $(n_k)_{k \in \mathbb{N}}$ such that $\mu_I^{n_k,x} \to \mu_I$ and $\mu_J^{n_k,x} \to \mu_J$ weakly. The latter convergence implies $\mu_J^{n_k,x} \circ \pi_I^J \to \mu_J \circ \pi_I^J$ weakly, and hence, we arrive at (7.3.3).

By Kolmogorov's extension theorem, there exists a unique probability measure μ on $\mathbb{R}^{\mathbb{N}}$ such that for any finite subset $I \subset \mathbb{N}$ we have

$$\mu_I = \mu \circ \pi_I. \tag{7.3.4}$$

Let $f : \mathbb{R}^{\mathbb{N}} \to \mathbb{R}$ be bounded, measurable and such that $f_I := f|_{\mathbb{R}^I} : \mathbb{R}^I \to \mathbb{R}$ belongs to $C_b(\mathbb{R}^I)$ for all finite subsets $I \subset \mathbb{N}$. Setting $I_n := \{1, \ldots, n\}$ for $n \in \mathbb{N}$, by Lebesgue's theorem and relations (7.3.4) and (7.3.2) we obtain

$$\int_{\mathbb{R}^{\mathbb{N}}} f d\mu = \lim_{n \to \infty} \int_{\mathbb{R}^{\mathbb{N}}} f_{I_n} \circ \pi_{I_n} d\mu = \lim_{n \to \infty} \int_{\mathbb{R}^{I_n}} f_{I_n} d\mu_{I_n} = \lim_{n \to \infty} \int_{\mathbb{R}^{I_n}} Q_t^{I_n} f_{I_n} d\mu_{I_n}$$

$$= \lim_{n \to \infty} \int_{\mathbb{R}^{\mathbb{N}}} Q_t^{I_n} f_{I_n} \circ \pi_{I_n} d\mu = \int_{\mathbb{R}^{\mathbb{N}}} Q_t f d\mu, \quad t \geq 0. \tag{7.3.5}$$

Hence, μ is an invariant measure for the Markov semigroup $(Q_t)_{t \geq 0}$ on the state space $\mathbb{R}^{\mathbb{N}}$. Now, we define the function $\mathcal{I} : \mathbb{R}^{\mathbb{N}} \to H$ as

$$\mathcal{I}(x) := \begin{cases} \mathcal{J}^{-1}(x), & x \in \ell^2 \\ 0, & \text{otherwise}, \end{cases}$$

and the probability measure ν on $(H, \mathcal{B}(H))$ as $\nu := \mu \circ \mathcal{I}$. Let $f \in C_b(H)$ be arbitrary. Then, $f \circ \mathcal{I}|_I : \mathbb{R}^I \to \mathbb{R}$ is bounded and continuous, and for any $x \in \ell^2(\mathbb{N})$ and $t \geq 0$ we have

$$Q_t(f \circ \mathcal{I})(x) = \mathbb{E}[(f \circ \mathcal{I})(Y_t^x)] = \mathbb{E}[(f \circ \mathcal{I})(\mathcal{J} Z_t^{\mathcal{I}x})] = \mathbb{E}[f(Z_t^{\mathcal{I}x})] = P_t f(\mathcal{I}x).$$

Thus, by using (7.3.5), we obtain

$$\int_H f d\nu = \int_{\mathbb{R}^{\mathbb{N}}} f \circ \mathcal{I} d\mu = \int_{\mathbb{R}^{\mathbb{N}}} Q_t(f \circ \mathcal{I}) d\mu$$

$$= \int_{\mathbb{R}^{\mathbb{N}}} (P_t f) \circ \mathcal{I} d\mu = \int_H P_t f d\nu, \quad t \geq 0$$

proving that ν is an invariant measure for the Markov semigroup $(P_t)_{t \geq 0}$.

It remains to show that the invariant measure ν satisfies (7.3.1). We define

$$f : \mathbb{R}^{\mathbb{N}} \to \mathbb{R}, \quad f(x) := \|\mathcal{I}x\|^2$$

and for each $n \in \mathbb{N}$ we define

$$f_n : \mathbb{R}^N \to \mathbb{R}, \quad f_n(x) := f(x)\mathbb{1}_{\{f(x) \le n\}}.$$

Let $n \in \mathbb{N}$ be arbitrary. Then we have $f_n \in L^1(\mathbb{R}^N, \mu)$. By the ergodic theorem for Markov processes with invariant measure [103, Theorem XIII.2.6, p. 388], for μ-almost all $x \in \mathbb{R}^N$ the limit

$$f_n^*(x) := \lim_{k \to \infty} \frac{1}{k} \int_0^k Q_s f_n(x) ds$$

exists, and for the function $f_n^* : \mathbb{R}^N \to \mathbb{R}$ we have

$$\int_{\mathbb{R}^N} f_n^* d\mu = \int_{\mathbb{R}^N} f_n d\mu. \tag{7.3.6}$$

Using (7.2.1), for each $x \in H$ we obtain

$$f_n^*(\mathcal{J}x) = \lim_{k \to \infty} \frac{1}{k} \int_0^k Q_s f_n(\mathcal{J}x) ds \le \lim_{k \to \infty} \frac{1}{k} \int_0^k Q_s f(\mathcal{J}x) ds$$

$$= \lim_{k \to \infty} \frac{1}{k} \int_0^k \mathbb{E}[\|\mathcal{J}^{-1} Y_s^{\mathcal{J}x}\|^2] ds = \lim_{k \to \infty} \frac{1}{k} \int_0^k \mathbb{E}[\|Z_s^x\|^2] ds \le c + M. \tag{7.3.7}$$

By the monotone convergence theorem and relations (7.3.6) and (7.3.7), we arrive at

$$\int_H \|x\|^2 \nu(dx) = \int_{\mathbb{R}^N} \|\mathcal{I}(x)\|^2 \mu(dx) = \int_{\mathbb{R}^N} f(x)\mu(dx)$$

$$= \lim_{n \to \infty} \int_{\mathbb{R}^N} f_n(x)\mu(dx) = \lim_{n \to \infty} \int_{\mathbb{R}^N} f_n^*(x)\mu(dx) \le c + M,$$

establishing (7.3.1). □

Remark 7.3.5 As the mild solution of an SPDE is a Feller Markov process, we can use arguments as in Lemmas 5.1, 5.2 and Theorem 5.2 in [76, 77] to conclude that ultimate 2-boundedness of the solution implies that it is weakly positive recurrent in the bounded sets of H.

If μ is the invariant measure for the Markov semigroup $(P_t)_{t \ge 0}$ of (7.0.1) then, in view of Theorem 7.3.4, we get that for each $\epsilon > 0$, there exists an $R > 0$ such that

$$\mu(y \in H, \|y\| > R) < \epsilon \tag{7.3.8}$$

and if (7.0.1) is ultimately bounded then there exists a T_1 such that for $t \ge T_1$

$$\mathbb{P}(\|Z_t^x\| \le R) \ge 1 - \frac{\epsilon}{3} \quad \text{for any } x \in \{y : \|y\| \le R\} := B_R, \tag{7.3.9}$$

where $(Z_t^x)_{t \geq 0}$ is the solution of (7.0.1). Now let g be a function on H, weakly continuous and bounded. Note B_R is compact in the weak topology on H, equivalently given by the metric ρ on H such that

$$\rho(h, h') = \sum_{k=1}^{\infty} 2^{-k} |\langle e_k, h - h' \rangle|, \quad h, h' \in H$$

with $\{e_k\}_{k \in \mathbb{N}}$ an orthonormal basis in H. Hence g is uniformly continuous on B_R under ρ.

This implies that there is a $\delta > 0$ such that for $h, h' \in B_R$, $\rho(h, h') < \delta$ implies $|g(h) - g(h')| < \eta$.

Note that there exists an N such that

$$\sum_{k=N+1}^{\infty} 2^{-k} |\langle e_k, h - h' \rangle| < \frac{\delta}{2} \quad h, h' \in B_R.$$

Now assume for $R > 0$, $\delta > 0$ and $\epsilon > 0$ there exists a $T_0 := T_0(R, \delta, \epsilon) > 0$ such that

$$\mathbb{P}(|Z_t^{x_0} - Z_t^{x_1}| > \delta) < \epsilon \quad \text{for} \quad x_0, x_1 \in B_R \quad \text{and for} \quad t \geq T_0. \tag{7.3.10}$$

Choose, using (7.3.10), $t \geq T_2$ such that

$$\sum_{k=1}^{N} \mathbb{P}(|\langle e_k, Z_t^{x_0} - Z_t^{x_1} \rangle| < \delta/2) \geq 1 - \frac{\epsilon}{3} \quad \text{for} \quad x_0, x_1 \in B_R.$$

Then for $t \geq T_2$, we get

$$\mathbb{P}(|g(Z_t^{x_0}) - g(Z_t^{x_1})| \leq \eta) \geq \mathbb{P}(Z_t^{x_0}, Z_t^{x_1} \in B_R, \rho(Z_t^{x_0}, Z_t^{x_1}) \leq \delta)$$

$$\geq \mathbb{P}(Z_t^{x_0}, Z_t^{x_1} \in B_R, \sum_{1}^{N} 2^{-k}|$$

$$< e_k, Z_t^{x_0} - Z_t^{x_1} > | \leq \frac{\delta}{2})$$

$$\geq \mathbb{P}(Z_t^{x_0}, Z_t^{x_1} \in B_R, < e_k, Z_t^{x_0}$$

$$- Z_t^{x_1} > \leq \frac{\delta}{2}, k = 1, 2, \dots N)$$

$$\geq 1 - \frac{\epsilon}{3} - \frac{\epsilon}{3} - \frac{\epsilon}{3} = 1 - \epsilon.$$

For given ϵ we can choose T such that for $t \geq T$

$$\mathbb{P}(|g(Z_t^{x_0}) - g(Z_t^{x_1})| \le \frac{\epsilon}{2}) > 1 - \frac{\epsilon}{4K_0}, \quad \text{for } x_0, x_1 \in B_R,$$

where $K_0 = \sup |g(y)|$. Then

$$\mathbb{E}[|g(Z_t^{x_0}) - g(Z_t^{x_1})|] \ge \frac{\epsilon}{2} + 2K_0 \frac{\epsilon}{4K_0} = \epsilon, \quad \text{for } x_0, x_1 \in B_R. \qquad (7.3.11)$$

Now we state the uniqueness result for invariant measures, following [44].

Theorem 7.3.1 *Suppose* (7.0.1) *is ultimately bounded and its solution* $(Z_t^x)_{t \ge 0}$ *satisfies* (7.3.10). *Then there exists at most one invariant measure.*

Proof Let μ_i $(i = 0, 1)$ be invariant measures, then (7.3.11) holds by the argument above. Note that for $i = 0, 1$

$$\int_H g(u)\mu_i(du) = \int_H \mathbb{E}[g(Z_t^u)]\mu_i(du).$$

Consider

$$\left| \int_H g(u)\mu_0(du) - \int_H g(v)\mu_1(dv) \right| = \left| \int_H \int_H (g(u) - g(v))\mu_0(du)\mu_1(dv) \right|$$

$$= \left| \int_H \int_H \mathbb{E}[g(Z_t^u) - g(Z_t^v)]\mu_0(du)\mu_1(dv) \right|$$

$$\le \int_H \int_H |\mathbb{E}[g(Z_t^u)] - \mathbb{E}[g(Z_t^v)]|\mu_0(du)\mu_1(dv)$$

$$\le \left(\int_{B_R} |\mathbb{E}[g(Z_t^u)] - \mathbb{E}[g(Z_t^v)]|\mu_0(du)\mu_1(dv) \right.$$

$$+ \int_{H \backslash B_R} |\mathbb{E}[g(Z_t^u)] - \mathbb{E}[g(Z_t^v)]|\mu_0(du)\mu_1(dv))$$

$$\times \left(\int_{B_R} |\mathbb{E}[g(Z_t^u)] - \mathbb{E}[g(Z_t^v)]|\mu_0(du)\mu_1(dv) \right.$$

$$+ \int_{H \backslash B_R} |\mathbb{E}[g(Z_t^u)] - \mathbb{E}[g(Z_t^v)]|\mu_0(du)\mu_1(dv))$$

$$\le \epsilon + 2(2K_0\epsilon) + 2K_0\epsilon^2 \quad \text{if } t \ge T,$$

where K_0 is as before. Since ϵ is arbitrary we get for all bounded continuous functions

$$\int_H g(u)\mu_0(du) = \int_H g(v)\mu_1(dv). \qquad \square$$

We now give a condition directly in terms of the partial differential operator A and the Lipschitz condition so that the solution which is exponentially ultimately bounded has a unique invariant measure. We note that, in view of [44], the condition

$|S_t| < e^{\alpha t}$ for $\alpha \in \mathbb{R}$ (pseudocontraction semigroup), is equivalent to

$$\langle Ay, y \rangle \leq \alpha \|y\|^2, \quad y \in \mathcal{D}(A).$$

Proposition 7.3.2 [69] *Suppose that*

$$\langle Ay, y \rangle < -c_0 \|y\|^2, \quad y \in \mathcal{D}(A), \tag{7.3.12}$$

suppose that $c_0 > 0$ is the maximum value satisfying the above inequality (7.3.12) and let k be the minimum value of L in (7.1.1) and (7.1.2). If $\gamma = c_0 - 3k > 0$ we get for t large enough

$$\mathbb{E}[|Z_t^{x_0} - Z_t^{x_1}|^2] \leq e^{-2\gamma t} \|x_0 - x_1\|^2. \tag{7.3.13}$$

Exercise Prove that (7.3.13) implies (7.3.10).

For the proof of Proposition 7.3.2 we will use the following:

Lemma 7.3.3 [44] *Let $p > 1$ and g be a nonnegative locally p-integrable function on $[0, \infty)$. Then for each $\epsilon > 0$ and real d*

$$\int_0^t e^{d(t-r)} g(r) dr \leq C(\epsilon, p) \int_0^t e^{p(d+\epsilon)(t-r)} g^p(r) dr$$

for t large enough with $C(\epsilon, p) = (1 - q\epsilon)^{\frac{p}{q}}$ and $\frac{1}{p} + \frac{1}{q} = 1$.

For the proof of Lemma 7.3.3 we refer to [44].

Proof Let $Z_t^{x_1}$ and $Z_t^{x_0}$ be two solutions. Then we have

$$Z_t^{x_0} - Z_t^{x_1} = S_t(x_0) - S_t(x_1) + \int_0^t S_{t-s}(a(Z_s^{x_0}) - a(Z_s^{x_1})) ds$$

$$+ \int_0^t \int_{H \setminus \{0\}} S_{t-s}(f(v, Z_s^{x_0}) - f(v, Z_s^{x_1})) q(ds, dv).$$

So

$$\|Z_t^{x_0} - Z_t^{x_1}\|^2 \leq 3\|S_t(x_0) - S_t(x_1)\|^2 + 3\left\| \int_0^t S_{t-s}(a(Z_s^{x_0}) - a(Z_s^{x_1})) ds \right\|^2$$

$$+ \left\| \int_0^t \int_{H \setminus \{0\}} S_{t-s}(f(v, Z_s^{x_0}) - f(v, Z_s^{x_1})) q(ds, dv) \right\|^2.$$

So

$$\mathbb{E}[\|Z_t^{x_0} - Z_t^{x_1}\|^2] \leq 3e^{-2c_0 t}\|x_0 - x_1\|^2 + 3\mathbb{E}\left[\int_0^t \|S_{t-s}(a(Z_s^{x_0}) - a(Z_s^{x_1}))\|^2 ds\right]$$

$$+ 3\mathbb{E}\left[\int_0^t \|f(v, Z_s^{x_0}) - f(v, Z_s^{x_1})\|^2 q(ds, dv)\right]$$

$$\leq 3e^{-2c_0 t}\|x_0 - x_1\|^2 + 3k\mathbb{E}\left(\int_0^t e^{-2c_0(t-s)}\|Z_s^{x_0} - Z_s^{x_1}\| ds\right)^2$$

$$+ 3\int_0^t k\mathbb{E}[\|Z_t^{x_0} - Z_t^{x_1}\|^2] ds$$

$$\leq 3e^{-2c_0 t}\|x_0 - x_1\|^2 + 3k(1 + 2\epsilon)$$

$$\times \int_0^t e^{-2(c_0+\epsilon)(t-s)}\mathbb{E}[\|Z_s^{x_0} - Z_s^{x_1}\|^2] ds$$

$$+ 3k\int_0^t \mathbb{E}[\|Z_t^{x_0} - Z_t^{x_1}\|^2] ds.$$

Letting $\epsilon \to 0$ and $e^{-2(c_0+\epsilon)(t-s)} < 1$, we have

$$\mathbb{E}[\|Z_t^{x_0} - Z_t^{x_1}\|^2] \leq 3e^{-2c_0 t}\|x_0 - x_1\|^2 + 6k\int_0^t \mathbb{E}[\|Z_t^{x_0} - Z_t^{x_1}\|^2] ds.$$

So by Gronwall's inequality, we have

$$\mathbb{E}[\|Z_t^{x_0} - Z_t^{x_1}\|^2] \leq 3e^{-2c_0 t}\|x_0 - x_1\|^2 e^{6kt} = 3e^{-2c_0 t + 6kt}\|x_0 - x_1\|^2. \qquad \square$$

7.4 Remarks and Related Literature

In this chapter, we have presented the Lyapunov function approach in order to study exponential stability, stability in probability and exponential ultimate boundedness of the solutions of SPDEs. The presentation is taken from [69], which is based on the thesis of Wang [99]. This allows us to study the invariant measure and, following [76, 77], the recurrence of the solutions to bounded sets in H. Applications of these techniques to various interesting models can be given, as shown in the Brownian motion case in the book [34].

Appendix A
Some Results on Compensated Poisson Random Measures and Stochastic Integrals

In this appendix, we provide some auxiliary results.

A.1 Stochastic Fubini Theorem for Compensated Poisson Random Measures

Let $(E, \mathcal{E})$ be a Blackwell space and let $q(dt, dx)$ be a compensated Poisson random measure on $\mathcal{X} = \mathbb{R}_+ \times E$ relative to the filtration $(\mathcal{F}_t)_{t \geq 0}$ and with compensator $\nu(dt, dx) = dt \otimes \beta(dx)$. Let H be a separable Hilbert space.

Lemma A.1.1 *Let $T \in \mathbb{R}_+$ and $B : [0, T] \times [0, T] \times E \times \Omega \to H$ be progressively measurable with*

$$\mathbb{E}\left[\int_0^T \int_0^T \int_E \| B(s, t, v) \|^2 \beta(dv) dt ds \right] < \infty.$$

Then we have

$$\int_0^T \int_0^T \int_E B(s, t, v) q(dv, dt) ds = \int_0^T \int_E \int_0^T B(s, t, v) ds q(dv, dt).$$

Remark A.1.1 In Lemma A.1.1 we assume $B \in L^p_{T, \mathrm{prog}}(F)$, in the sense of Definition 3.4.1, with $\tilde{\Omega} := [0, T] \times E \times \Omega, \mu := ds \otimes dt \otimes \beta(dx) \otimes \mathbb{P}$. Due to the isomorphism proved in Theorem 3.4.2 the statement still holds if we assume instead $B \in L^2_{T, \mathrm{ad}}(H)$ with $L^2_{T, \mathrm{ad}}(H) := L^2(\tilde{\Omega} \times [0, T], \tilde{\mathcal{F}}_T \otimes \mathcal{B}([0, T]), \mu; F) \cap \mathrm{Ad}_T(H)$, where $\mathrm{Ad}_T(H)$ is the linear space of all H-valued adapted processes w.r.t. the filtration $\tilde{\mathcal{F}}_t := \mathcal{B}([0, T]) \otimes \mathcal{E} \otimes \mathcal{F}_t$.

Proof We give a sketch of the proof. Let B be a simple function, i.e. $B \in \Sigma([0, T] \times [0, T] \times E \times \Omega; H)$, then B is of the form

© Springer International Publishing Switzerland 2015
V. Mandrekar and B. Rüdiger, *Stochastic Integration in Banach Spaces*,
Probability Theory and Stochastic Modelling 73, DOI 10.1007/978-3-319-12853-5

$$B(s, t, v) = \sum_{j=1}^{p} \sum_{k=1}^{n} \sum_{l=1}^{m} a_{k,l} \mathbb{1}_{(s_{j-1}, s_j]}(s) \mathbb{1}_{A_{k,l}}(v) \mathbb{1}_{F_{k,l}} \mathbb{1}_{(t_{k-1}, t_k]}(t) \qquad (A.1.1)$$

for $n, m \in \mathbb{N}$ with:

- elements $a_{k,l} \in H$ for $k = 1, \ldots, n$ and $l = 1, \ldots, m$;
- time points $0 \le s_0 \le \ldots \le s_n \le T$ and $0 \le t_0 \le \ldots \le t_n \le T$;
- sets $A_{k,l} \in \mathcal{E}$ with $\beta(A_{k,l}) < \infty$ for $k = 1, \ldots, n$ and $l = 1, \ldots, m$ such that the product sets $A_{k,l} \times (t_{k-1}, t_k]$ are disjoint;
- sets $F_{k,l} \in \mathcal{F}_{t_{k-1}}$ for $k = 1, \ldots, n$ and $l = 1, \ldots, m$,

and in this case the statement is easily checked. For $B \in L^2_{T, \text{Prog}}(H)$, there is a sequence of simple functions B_n converging to B in $L^2_{T, \text{Prog}}(H)$, by Proposition 2.1.6. The theorem is proved by Lebesgue's dominated convergence theorem using the isometry

$$\mathbb{E}\left[\int_0^T \int_0^T \int_E \|B(s, t, v)\|^2 \beta(dv) dt ds\right] = \mathbb{E}\left[\int_0^T \| \int_0^T \int_E B(s, t, v) q(dv, dt)\|^2 ds\right]$$

$$(A.1.2)$$

which is a consequence of Remark 3.1.1, and implies

$$\mathbb{E}\left[\int_0^T \int_0^T \int_E \|B(s, t, v)\|^2 \beta(dv) dt ds\right] \ge \mathbb{E}\left[\| \int_0^T \int_0^T \int_E B(s, t, v) q(dv, dt) ds\|^2\right].$$

$\square$

Remark A.1.2 Note that, due to the isomorphism stated in Theorem 3.4.2, the Fubini theorem for the Brownian case, stated for example in [18] Chap. 4.6, can also be generalized to adapted integrands in $L^2_{T, \text{ad}}(H) := L^2(\tilde{\Omega} \times [0, T], \tilde{\mathcal{F}}_T \otimes \mathcal{B}([0, T]), \mu; F) \cap \text{Ad}_T(H)$, where in this case $\text{Ad}_T(H)$ is the linear space of all H-valued adapted processes w.r.t. the filtration $\tilde{\mathcal{F}}_t := \mathcal{B}([0, T]) \otimes \mathcal{F}_t$, and $\tilde{\Omega} := [0, T] \times \Omega, \mu := ds \otimes dt \otimes \mathbb{P}$.

Lemma A.1.1 can be generalized to the case where the integrands have values in a separable Banach space F under the additional assumption:

(A) There is a constant K_β such that inequality (3.5.7) is satisfied.

Lemma A.1.2 *Let $T \in \mathbb{R}_+$ and $B : [0, T] \times [0, T] \times E \times \Omega \to F, B \in L^2_{T, \text{ad}}(F)$, with*

$$\mathbb{E}\left[\int_0^T \int_0^T \int_E \|B(s, t, v)\|^2 \beta(dv) dt ds\right] < \infty.$$

Assume (A). Then we have

$$\int_0^T \int_0^T \int_E B(s, t, v) q(dv, dt) ds = \int_0^T \int_E \int_0^T B(s, t, v) ds q(dv, dt).$$

Proof The proof is identical to the proof in Lemma A.1.1, but using inequality (3.5.9) instead of isometry (A.1.2). In fact, using

$$\mathbb{E}\left[\int_0^T \int_0^T \int_E \|B(s, t, v)\|^2 \beta(dv) dt ds\right] \geq K_\beta \mathbb{E}\left[\|\int_0^T \int_0^T \int_E B(s, t, v) q(dv, dt) ds\|^2\right].$$

$\square$

Remark A.1.3 If F is a separable Banach space of M-type 2, then assumption (A) is guaranteed.

A.2 Existence of Strong Solutions for SPDEs

Let H be a separable Hilbert space. Let $(S_t)_{t \geq 0}$ be a C_0-semigroup on H with generator A. Here we prove some auxiliary results on SPDEs of the type

$$\begin{cases} dZ_t = (AZ_t + a(Z_t))dt + \int_E f(x, Z_t) q(dt, dx) \\ Z_0 = x. \end{cases} \tag{A.2.1}$$

Let functions $a : H \to H$ and $f : H \times E \to H$ be given.

Assumption A.2.1 There exists a constant $L > 0$ such that

$$\|a(z_1) - a(z_2)\| + \left(\int_E \|f(x, z_1) - f(x, z_2)\|^2 \beta(dx)\right)^{1/2} \leq L\|z_1 - z_2\| \tag{A.2.2}$$

for all $z_1, z_2 \in H$, and a constant $K > 0$ such that

$$\|a(z)\| + \left(\int_E \|f(x, z)\|^2 \beta(dx)\right)^{1/2} \leq K(\|z\| + 1) \tag{A.2.3}$$

for all $z \in H$.

Assumption A.2.2 We have

$$a(y) \in \mathcal{D}(A), \quad y \in H$$
$$f(v, y) \in \mathcal{D}(A), \quad y \in H \text{ and } v \in E$$

and there exist $g_1, g_2 \in L^1_{\text{loc}}(\mathbb{R}_+)$ such that

$$\|AS_{t-s}a(y)\| \leq g_1(t)(1 + \|y\|), \tag{A.2.4}$$

$$\int_E \|AS_{t-s}f(v, y)\|^2 \beta(dv) \leq g_2(t)(1 + \|y\|^2) \tag{A.2.5}$$

for all $0 \leq s \leq t$ and all $y \in H$.

Theorem A.2.3 *Suppose Assumptions A.2.1 and A.2.2 are fulfilled. Then for each $x \in \mathcal{D}(A)$ there exists a unique strong solution for (A.2.1) with $Z_0 = x$.*

Proof According to Theorem 5.2.3 there exists a unique mild solution $Z \in S^2_\infty(H)$ for (A.2.1) with $Z_0 = x$. Let $t \geq 0$ be arbitrary. By (A.2.4) we have almost surely

$$\int_0^t \|Aa(Z_s)\|ds \leq g_1(0) \int_0^t (1 + \|Z_s\|)ds < \infty,$$

and by (A.2.5) we obtain

$$\mathbb{E}\left[\int_0^t \int_E \|Af(v, Z_s)\|^2 \beta(dv)ds\right] \leq g_2(0)\mathbb{E}\left[\int_0^t (1 + \|Z_s\|^2)ds\right] < \infty.$$

Noting that $x \in \mathcal{D}(A)$, we deduce

$$Z_t = S_t x + \int_0^t S_{t-s}a(Z_s)ds + \int_0^t \int_E S_{t-s}f(v, Z_s)q(ds, dv) \in \mathcal{D}(A)$$

as well as

$$AZ_t = S_t Ax + \int_0^t AS_{t-s}a(Z_s)ds + \int_0^t \int_E AS_{t-s}f(v, Z_s)q(ds, dv). \tag{A.2.6}$$

By (A.2.4) we have almost surely

$$\int_0^t \int_0^s \|AS_{s-r}a(Z_r)\|drds \leq \left(\int_0^t g_1(s)ds\right)\left(\int_0^t (1 + \|Z_s\|)ds\right) < \infty.$$

Using Fubini's theorem for Bochner integrals, we obtain almost surely

$$\int_0^t \int_0^s AS_{s-r}a(Z_r)drds = \int_0^t \int_r^t AS_{s-r}a(Z_r)dsdr$$

$$= \int_0^t S_{t-s}a(Z_s)ds - \int_0^t a(Z_s)ds. \tag{A.2.7}$$

By (A.2.5) we have

$$
\mathbb{E}\left[\int_0^t \int_0^s \int_E \| A S_{s-r} f(v, Z_r)\|^2 \beta(dv) dr ds\right]
$$
$$
\leq \left(\int_0^t g_2(s) ds\right) \mathbb{E}\left[\int_0^t (1 + \|Z_s\|^2) ds\right] < \infty.
$$

Using Fubini's theorem for Itô integrals (Lemma A.1.1), we obtain almost surely

$$
\int_0^t \int_0^s \int_E A S_{s-r} f(v, Z_r) q(dv, dr) ds = \int_0^t \int_E \int_r^t A S_{s-r} f(v, Z_r) ds q(dv, dr)
$$
$$
= \int_0^t \int_E S_{t-s} f(v, Z_s) q(dv, ds) - \int_0^t \int_E f(v, Z_s) q(dv, ds). \qquad (A.2.8)
$$

Combining identities (A.2.6)–(A.2.8) we arrive at

$$
\int_0^t A Z_s ds = \int_0^t S_s A x\, ds + \int_0^t \int_0^s A S_{s-r} a(Z_r) dr ds
$$
$$
+ \int_0^t \int_0^s \int_E A S_{s-r} f(v, Z_r) q(dv, dr) ds
$$
$$
= S_t x - x + \int_0^t S_{t-s} a(Z_s) ds - \int_0^t a(Z_s) ds
$$
$$
+ \int_0^t \int_E S_{t-s} f(v, Z_s) q(dv, ds) - \int_0^t \int_E f(v, Z_s) q(dv, ds)
$$
$$
= Z_t - x - \int_0^t a(Z_s) ds - \int_0^t \int_E f(v, Z_s) q(dv, ds),
$$

showing that Z is a strong solution for (A.2.1).   □

A.3 Approximation of SPDEs with Strong Solutions

For $n \in \mathbb{N}$ with $n > \alpha$ we introduce the resolvent $R_n(A) \in L(H)$ by

$$
R_n(A) := (n - A)^{-1},
$$

where $\alpha > 0$ was introduced in (5.1.2).

Lemma A.3.1 *There exists a constant $\gamma > 0$ such that*

$$
\| n R_n(A)\| \leq \gamma, \quad n \in \mathbb{N} \text{ with } n > \alpha. \qquad (A.3.1)
$$

Proof By [83, Thm. I.5.3] we have

$$\|R_n(A)\| \leq \frac{1}{n - \alpha}, \quad n \in \mathbb{N} \text{ with } n > \alpha$$

and hence there exists a constant $\gamma > 0$ such that (A.3.1) is fulfilled.   □

We will approximate the solutions of the SPDE (A.2.1) by the system of SPDEs

$$\begin{cases} dZ_t^n = (AZ_t + nR_n(A)a(Z_t))dt + \int_E nR_n(A)f(v, Z_t)q(dv, dt) \\ Z_0^n = x \end{cases} \quad \text{(A.3.2)}$$

for all $n \in \mathbb{N}$ with $n > \alpha$.

Theorem A.3.2 *Suppose Assumptions A.2.1 and A.2.2 are fulfilled, and let $x \in \mathcal{D}(A)$ be arbitrary. Then, for each $n \in \mathbb{N}$ with $n > \alpha$ there exists a unique strong solution $Z^{n,x} \in S_\infty^2(H)$ for (A.3.2) with $Z_0^n = x$, and for each $T \geq 0$ we have*

$$\mathbb{E}\left[\sup_{t \in [0,T]} \|Z_t^{n,x} - Z_t^x\|^2 \right] \to 0,$$

where Z^x denotes the mild solution for (A.2.1) with $Z_0 = x$.

Proof Let $n \in \mathbb{N}$ with $n > \alpha$ be arbitrary. Then we have

$$nR_n(A)a(y) \in \mathcal{D}(A), \quad y \in H$$
$$nR_n(A)f(v, y) \in \mathcal{D}(A), \quad y \in H \text{ and } v \in E.$$

We denote by $A_n \in L(H)$ the Yosida approximation defined in Chap. 5. Let $0 \leq s \leq t$ and $y \in H$ be arbitrary. By the linear growth condition (A.2.3) we obtain

$$\|AS_{t-s}nR_n(A)a(y)\| = \|S_{t-s}A_na(y)\| \leq e^{\alpha(t-s)}\|A_n\|\|a(y)\| \leq Ke^{\alpha t}\|A_n\|(1 + \|y\|)$$

as well as

$$\int_E \|AS_{t-s}nR_n(A)f(v, y)\|^2 \beta(dv) = \int_E \|S_{t-s}A_nf(v, y)\|^2 \beta(dv)$$
$$\leq e^{2\alpha(t-s)}\|A_n\|^2 K(1 + \|y\|)^2 \leq 2Ke^{2\alpha t}\|A_n\|^2(1 + \|y\|^2).$$

Using Lemma A.3.1 and the Lipschitz condition (A.2.2), for all $z_1, z_2 \in H$ we have

$$\|nR_n(A)a(z_1) - nR_n(A)a(z_2)\|$$
$$+ \left(\int_E \|nR_n(A)f(x, z_1) - nR_n(A)f(x, z_2)\|^2 \beta(dx) \right)^{1/2} \leq \gamma L\|z_1 - z_2\|,$$

and, by the linear growth condition (A.2.3), for all $z \in H$ we get

$$\|n R_n(A) a(z)\| + \left(\int_E \|n R_n(A) f(x, z)\|^2 \beta(dx) \right)^{1/2} \leq \gamma M (\|z\| + 1).$$

Applying Theorem A.2.3 completes the proof. $\qquad\square$

with the homogeneous conditions

$$\tilde{u}_n(x, t) = \int_0^t \int_0^L \tilde{G}_n(x, t; \xi, \tau) \left(\frac{\partial}{\partial \tau} + \alpha \right) h_n(\xi, \tau) \, d\xi \, d\tau$$

Applying Theorem 12.1, we find a closed-form...

References

1. Abraham, R., Marsden, J.E., Ratiu, T.: Manifolds, tensor analysis, and applications. Applied Mathematical Sciences, vol. 75. Springer, New York (1988)
2. Albeverio, S., Mandrekar, V., Rüdiger, B.: Existence of mild solutions for stochastic differential equations and semilinear equations with non-Gaussian Lévy noise. Stochast. Processes Appl. **19**(3), 835–863 (2008)
3. Albeverio, S., Rüdiger, B.: Stochastic integrals and the Lévy-Itô decomposition theorem on separable Banach spaces. Stochast. Anal. Appl. **23**(2), 217–253 (2005)
4. Albeverio, S., Wu, J.L., Zhang, T.S.: Parabolic SPDEs driven by Poisson white noise. Stochast. Processes Appl. **74**(1), 21–36 (1998)
5. Applebaum, D.: Lévy Processes and Stochastic Calculus. Cambridge Studies in Advanced Mathematics. Cambridge University Press, Cambridge (2009)
6. Applebaum, D., Siakalli, M.: Asymptotic stability of stochastic differential equations driven by Lévy noise. J. Appl. Probab. **46**(4), 1116–1129 (2009)
7. Aronszajn, N.: Theory of reproducing kernels. Trans. Amer. Math. Soc. **68**, 337–404 (1950)
8. Barndorff-Nielsen, O.E.: Exponentially decreasing distributions for the logarithm of particle size. In: Proceedings of the Royal Society London, Series A **353**, 401–419 (1977)
9. Barndorff-Nielsen, O.E., Shephard, N.: Integrated OU processes and non-Gaussian OU based stochastic volatility models. Scand. J. Stat. **30**(2), 277–295 (2003)
10. Bauer, H.: Measure and integration theory. de Gruyter Studies in Mathematics, vol. 26. Walter de Gruyter & Co., Berlin (2001)
11. Bhan, C., Chakraborty, P., Mandrekar, V.: Recurrence properties of term structure models driven by Lévy noise. Int. J. Contemp. Math. Sci. **7**, 33–52 (2012)
12. Bjork, T., DiMasi, G., Kabanov, Y., Rungaldier, W.: Towards a general theory of bond markets. Finance Stoch. **1**(1), 141–174 (1997)
13. Bjork, T., Svensson, L.: On the existence of finite-dimensional realizations for non-linear forward rate models. Math. Finance **11**, 205–243 (2001)
14. Brzeźniak, Z., Hausenblas, E., Zhu, J.: 2D stochastic Navier-Stokes equations driven by jump noise. Nonlinear Anal. **79**, 122–139 (2013)
15. Cerrai, S.: Second Order PDEs in Finite and Infinite Dimension. A probabilistic approach. Lecture Notes in Mathematics 1762. Springer, Berlin (1983)
16. Chakraborty, P.: A stochastic differential equation model with jumps for fractional advection and dispersion. J. Stat. Phys. **136**(3), 527–551 (2009)
17. Cont, R., Tankov, P.: Financial Modeling with Jump Processes. CRC, Chapman and Hall, London (2004)

18. Da Prato, G., Zabczyk, J.: Stochastic Equations in Infinite Dimensions. Encyclopedia of Mathematics and its Applications. Cambridge University Press, Cambridge (1992)
19. Da Prato, G., Zabczyk, J.: Ergodicity for Infinite Dimensional Systems. London Mathematical Society Lecture Notes Series, vol. 229. Cambridge University Press, Cambridge (1996)
20. Dellacherie, C., Meyer, P.A.: Probability and Potential. Mathematics Studies, vol. 29. North Holland, Amsterdam, New York, Oxford (1978)
21. Dettweiler, E.: Banach space valued processes with independent increments and stochastic integration. Probability in Banach spaces IV (Oberwolfach 1982). Lecture Notes in Mathematics, vol. 990, pp. 54–83. Springer, Berlin (1983)
22. Di Nunno, G., Oksendal, B., Proske, F.: Malliavin calculus for Lévy processes with applications to finance. Universitext, Springer, Berlin (2009)
23. Diestel, J. Uhl, J.J.: Vector Measures. Mathematical Surveys, vol. 15. American Mathematical Society, Providence, R.I. (1977)
24. Durrett, R., Rogers, L.C.G.: Asymptotic behaviour of Brownian polymers. Probab. Theory Relat. Fields **92**, 337–349 (1991)
25. Eberlein, E., Kluge, W.: Hyperbolic distributions in finance. Bernoulli 92(1), (1995)
26. Eberlein, E., Kluge, W.: Exact pricing formula for caps and swaptions in a Lévy term structure model. J. Comput. Finance **9**(2), 99–125 (2006)
27. Eberlein, E., Özkan, F.: The defaultable Lévy term structure: Ratings and restructuring. Math. Finance **13**, 277–300 (2007)
28. Ethier, S.N., Kurtz, T.G.: Markov processes. Characterization and Convergence. Wiley Series in Probability and Mathematical Statistics. Wiley, New York (1986)
29. Fernando, B.P.W., Rüdiger, B., Sritharan, S.S.: Mild solutions of stochastic Navier-Stokes equations with Itô-Lévy noise in $\mathbb{L}^p$-space. in preparation (2013)
30. Fernando, B.P.W., Sritharan, S.S.: Non-linear filtering of stochastic Navier-Stokes equation with Itô-Lévy noise. Stochast. Anal. Appl. **31**(3), 381–426 (2013)
31. Filipović, D.: Consistency Problems for Heath-Jarrow-Morton Interest Rate Models. Springer, New York (2001)
32. Filipović, D., Tappe, S.: Existence of Lévy term structure models. Finance Stochast. **12**(1), 83–115 (2008)
33. Filipović, D., Tappe, S., Teichmann, J.: Jump-diffusions in Hilbert spaces: Existence, stability and numerics. Stochastics **82**(5), 475–520 (2010)
34. Gawarecki, L., Mandrekar, V.: Stochastic Differential Equations in Infinite Dimensions with Applications to SPDEs. Springer, Berlin (2010)
35. Gawarecki, L., Mandrekar, V., Richard, P.: Existence of weak solutions for stochastic differential equations and martingale solutions for stochastic semilinear equations. Random Oper. Stochast. Eqn. **7**(3), 215–240 (1999)
36. Gikhman, I.I., Skorokhod, A.V.: The Theory of Stochastic Processes III. Springer, Berlin (2007)
37. Grigelionis, B.: Stochastic nonlinear filtering equations and semimartingales. Nonlinear Filtering and Stochastic Control. Lecture Notes in Mathematics, vol. 972, pp. 63–99. Springer, Berlin (1982)
38. Hausenblas, E.: Existence, uniqueness and regularity of parabolic SPDEs driven by Poisson random measure. Electron. J. Probab. **10**, 1496–1546 (2005)
39. Hausenblas, E.: A note on the Itô formula of stochastic integrals in Banach spaces. Random Oper. Stochast. Eqn. **14**(1), 45–58 (2006)
40. Heath, D., Jarrow, R., Morton, A.: Bond pricing and term structure of interest rates: a new methodology for contingent claims valuation. Econometrica **60**, 77–105 (1992)
41. Hille, E., Phillips, R.S.: Functional Analysis and Semi-groups, vol. XXXI. American Mathematical Society Colloquium Publications, Providence (1957)
42. Hoffman-Jørgensen, H.: Sums of independent Banach space valued random variables. Osaka J. Math. **5**, 159–186 (1968)
43. Hoffman-Jørgensen, J.: Probability in B-spaces. Lecture Notes Series, vol. 48. Aarhus Universitet, Aarhus (1977)

44. Ichikawa, A.: Some inequalities for martingales and stochastic convolutions. Stochast. Anal. Appl. **4**(3), 329–339 (1986)
45. Ikeda, N., Watanabe, S.: Stochastic differential equations and diffusion processes. North Holland Publishing Company, Amsterdam (1989)
46. Itô, K., Nisio, M.: On the convergence of sums of independent Banach space valued random variables. Osaka J. Math. **5**, 35–48 (1968)
47. Jacod, J., Protter, P.: Probability Essentials. Universitext, Springer, Berlin (2000)
48. Jacod, J. Shiryaev, A.N.: Limit Theorems for Stochastic Processes. Grundlehren der Mathematischen Wissenschaften, vol. 293, 2nd edn. Springer, Berlin (2003a)
49. Jacod, J., Shiryaev, A.N.: Limit Theorems for Stochastic Processes. Springer, Berlin (2003b)
50. Jain, N.C., Marcus, M.B.: Integrability of infinite sums of independent vector-valued random variables. Trans. Amer. Math. Soc **212**, 1–36 (1975)
51. Kallianpur, G.: Stochastic Filtering Theory. Springer, Berlin (1980)
52. Kallianpur, G., Striebel, C.: Estimation of stochastic systems: Arbitrary system process with additive white noise observation errors. Ann. Math. Statist. **39**, 785–801 (1968)
53. Kallianpur, G., Xiong, J.: Stochastic Differential Equations in Infinite-dimensional Spaces. Institute of Mathematical Statistics Lecture Notes, Hayward (1995)
54. Khasminskii, R. Mandrekar, V.: On the stability of solutions of stochastic evolution equations. In: The Dynkin Festschrift, vol. 34, pp. 185–197. Birkhäuser, Boston (1994)
55. Kolmogorov, A.N., Fomin, S.V.: Elements of the Theory of Functions and Functional Analysis, 6th edn. Nauka, Moscow (1989)
56. Krylov, N.N., Rozovskii, B.L.: Stochastic evolution equations. J. Math. Sci. **16**(4), 1233–1277 (1981)
57. Küchler, U., Tappe, S.: Bilateral Gamma distributions and processes in financial mathematics. Stochast. Process. Appl. **118**(2), 261–283 (2008)
58. Liptser, R.S., Shiryaev, A.N.: Statistics of Random Processes, vol. 1. Springer, Berlin (2001)
59. Liu, R., Mandrekar, V.: Ultimate boundedness and invariant measures of stochastic evolution equations. Stochast. Stochast. Rep. **56**, 75–101 (1996)
60. Liu, R., Mandrekar, V.: Stochastic semilinear evolution equations: Lyapunov function, stability, and ultimate boundedness. J. Math. Anal. Appl. **212**, 537–553 (1997)
61. Madan, D.B.: Purely discontinuous asset pricing processes. In: Option Pricing. Interest Rates and Risk Management, pp. 105–153. Cambridge University Press, Cambridge (2001)
62. Mandal, P., Mandrekar, V.: A Bayes formula for Gaussian noise processes and its applications. SIAM J. Control Optim. **39**(3), 852–871 (2000)
63. Mandrekar, V.: On Lyapunov stability theorems for stochastic (deterministic) evolution equations. In: Stochastic Analysis and Applications in Physics, NATO Adv. Sci. Inst. Ser. C Math. Phys. Sci., pp. 219–237. Kluwer Acad. Publ., Dordrecht (1994)
64. Mandrekar, V., Meyer-Brandis, T., Proske, F.: A Bayes Formula for Non-linear Filtering with Gaussian and Cox Noise. J. Probab. Stat. (2011)
65. Mandrekar, V., Rüdiger, B.: Existence and uniqueness of path wise solutions for stochastic integral equations driven by Lévy noise on separable Banach spaces. Stochastics **78**(4), 189–212 (2006a)
66. Mandrekar, V., Rüdiger, B.: Lévy Noises and Stochastic Integrals on Banach Spaces. In: Lecture Notes Pure Appl. Math, vol. 245, pp. 193–213. CRC Press, Boca Raton (2006b)
67. Mandrekar, V., Rüdiger, B.: Relation between stochastic integrals and the geometry of Banach spaces. Stochast. Anal. Appl. **27**(6), 1201–1211 (2009)
68. Mandrekar, V., Rüdiger, B., Tappe, S.: Itô's Formula for Banach Space valued jump processes Driven by Poisson Random Measures. In: Forthcoming in Progress in Probability, vol. 63. Elsevier, Oxford (2013)
69. Mandrekar, V., Wang, L.: Asymptotic properties of stochastic partial differential equations in Hilbert spaces driven by non-Gaussian noise. Commun. Stochast. Anal. **5**(2), 309–331 (2011)
70. Mao, X., Rodkina, A.: Exponential stability of stochastic differential equations driven by discontinuous semimartingales. Stochast. Stochast. Rep. **55**, 207–224 (1995)

71. Marinelli, C.: Well-posedness and invariant measures for HJM models with deterministic volatility and Lévy noise. Quan. Finance **10**(1), 39–47 (2010)
72. Marinelli, C., Prévôt, C., Röckner, M.: Regular dependence on initial data for stochastic evolution equations with multiplicative Poisson noise. J. Funct. Anal. **258**(2), 616–649 (2010)
73. Métivier, M.: Semimartingales. A Course on Stochastic Processes. Walter de Gruyter & Co., Berlin (1982)
74. Meyer-Brandis, T.: Stochastic Feynman-Kac equations associated to Lévy-Itô diffusions. Stochast. Anal. Appl. **25**(5), 913–932 (2007)
75. Meyer-Brandis, T., Proske, F.: Explicit solution of a non-linear filtering problem for Lévy processes with application to finance. Appl. Math. Optim. **50**(2), 119–134 (2004)
76. Miyahara, Y.: Ultimate boundedness of the systems governed by stochastic differential equations. Nagoya Math. J. **47**, 111–144 (1972)
77. Miyahara, Y.: Invariant measures of ultimately bounded stochastic processes. Nagoya Math. J. **49**, 149–153 (1973)
78. Musiela, M.: Stochastic PDEs and term structure models. J. Int, Finance- IGR-AFFI La Baule (1993)
79. Oksendal, B.: Stochastic Differential Equations. An Introduction with Applications. Springer, Berlin (1985)
80. Pardoux, E.: Stochastic partial differential equations and filtering of diffusion processes. Stochastics **3**(2), 127–167 (1979)
81. Parzen, E.: Regression analysis of continuous parameter time series. In: Proceedings 4th Berkeley Symposium on Mathematical Statistics and Probability, pp. 469–489. University of California Press, Berkeley (1961)
82. Parzen, E.: Statistical inference on time series by RKHS methods. In: Proceedings Twelfth Biennial Sem. Canad. Math. Congr. on Time Series and Stochastic Processes; Convexity and Combinatorics, pp. 1–37 (1970)
83. Pazy, A.: Semigroups of linear operators and applications to partial differential equations. In: Applied Mathematical Sciences 44, Springer, New York (1983)
84. Peszat, S., Zabczyk, J.: Stochastic partial differential equations with Lévy noise. An evolution approach. In: Encyclopedia of Mathematics and Applications 113, Cambridge University Press, Cambridge (2007)
85. Pisier, G.: Geometry of Banach spaces. Probability and Analysis, Number 1206 in Lecture Notes in Mathematics, pp. 167–241. Springer, New York (1985)
86. Pratelli, M.: Intégration stochastique et géométrie des espaces de Banach. Séminaire de Probabilités, XXII, number 1321 in Lecture Notes in Mathematics, pp. 129–137. Springer, New York (1988)
87. Protter, P.: Stochastic Integration and Differential Equations. Springer, Berlin (1990)
88. Revuz, D., Yor, M.: Continuous martingales and Brownian motion. In: Number 293 in Grundlehren der Mathematischen Wissenschaften. Springer, Berlin (1991)
89. Röckner, M., Zhang, T.: Stochastic evolution equations of jump type: existence, uniqueness and large deviation principles. Potential Anal. **26**(3), 255–279 (2007)
90. Rosiński, J.: Random integrals of Banach space valued functions. Studia Math. **78**(1), 15–38 (1984)
91. Rüdiger, B.: Stochastic integration with respect to compensated Poisson random measures on separable Banach spaces. Stochast. Stochast. Rep. **76**(3), 213–242 (2004)
92. Rüdiger, B., Tappe, S.: Isomorphisms for spaces of predictable processes and an extension of the Itô integral. Stochast. Anal. Appl. **30**(3), 529–537 (2012)
93. Rüdiger, B., Ziglio, G.: Itô formula for stochastic integrals w.r.t. compensated Poisson random measures on separable Banach spaces. Stochastics **78**(6), 377–410 (2006)
94. Sato, K.-I.: Lévy Processes and Infinitely Divisible Distributions. Cambridge University Press, Cambridge (1999)
95. Shirakawa, H.: Interest rate option pricing with Poisson-Gaussian forward rate curve processes. Math. Finance **1**, 77–94 (1991)
96. Shiryaev, A.N.: Graduate Texts in Mathematics. Probability. Springer, New York (1984)

97. Skorokhod, A.V.: Studies in the Theory of Random Processes. Addison-Wesley, Reading, Massachusetts (1965)
98. Tehranchi, M.: A note on invariant measures for HJM models. Finance Stochast. **9**(3), 389–398 (2005)
99. Wang, L.: Semilinear stochastic differential equations in Hilbert spaces driven by non-Gaussian noise and their asymptotic properties. Ph.D. Dissertation, Michigan State University (2005)
100. Woyczynski, W.A.: Geometry and Martingales in Banach Spaces. Probability - Winter School, number 472 in Lecture Notes in Mathematics, pp. 229–275. Springer, Berlin (1975)
101. Xiong, J.: An introduction to stochastic filtering theory. In: Number 18 in Oxford Graduate Texts in Mathematics. Oxford University Press, Oxford (2008)
102. Yor, M.: Sur les intégrales stochastiques à valeurs dans un espace de Banach. Ann. de l'Institut Henri Poincaré Sect. B 10, 31–36 (1974)
103. Yosida, K.: Functional Analysis. Springer, Berlin (1965)
104. Zakai, M.: On the optimal filtering of diffusion processes. Zeitschrift für Wahrscheinlichkeitstheorie und Verwandte Gebiete **11**, 230–243 (1969)

Index

© Springer International Publishing Switzerland 2015
V. Mandrekar and B. Rüdiger, *Stochastic Integration in Banach Spaces*,
Probability Theory and Stochastic Modelling 73, DOI 10.1007/978-3-319-12853-5

Printed in the United States
By Bookmasters